Classical Fluid Mechanics

Authored by

Michael Belevich
St.Petersburg
Russia

General:

1. Any dispute or claim arising out of or in connection with this License Agreement or the Work (including non-contractual disputes or claims) will be governed by and construed in accordance with the laws of the U.A.E. as applied in the Emirate of Dubai. Each party agrees that the courts of the Emirate of Dubai shall have exclusive jurisdiction to settle any dispute or claim arising out of or in connection with this License Agreement or the Work (including non-contractual disputes or claims).
2. Your rights under this License Agreement will automatically terminate without notice and without the need for a court order if at any point you breach any terms of this License Agreement. In no event will any delay or failure by Bentham Science Publishers in enforcing your compliance with this License Agreement constitute a waiver of any of its rights.
3. You acknowledge that you have read this License Agreement, and agree to be bound by its terms and conditions. To the extent that any other terms and conditions presented on any website of Bentham Science Publishers conflict with, or are inconsistent with, the terms and conditions set out in this License Agreement, you acknowledge that the terms and conditions set out in this License Agreement shall prevail.

Bentham Science Publishers Ltd.
Executive Suite Y - 2
PO Box 7917, Saif Zone
Sharjah, U.A.E.
Email: subscriptions@benthamscience.org

BENTHAM SCIENCE

CONTENTS

FOREWORD

Writing a new fluid dynamics textbook is a challenging task. In 1895, Sir Horace Lamb established a very high standard with the first edition of Hydrodynamics. This classical presentation was followed by other excellent introductions into the field of fluid mechanics, among them Landau and Lifshitz, 1959, and Batchelor, 1967. The strength of M.Belevich's book is in its rigorous and systematic approach to developing the mathematical model of fluid dynamics from the first principles. It carefully explains the underlying hypothesis and simplifications used to establish equations that govern motions of a fluid. Extensive use of vector and tensor analysis results in a compact and generalized narrative, without the restrictions of a particular coordinate system.

This textbook is by no means a comprehensive description of the field of fluid dynamics. Some of important problems (*e.g.* waves) were deliberately left out of the book's framework. Since the text is based on a course that is taught to students who specialize in geophysical fluid dynamics, more engineering aspects of fluid mechanics (such as turbomachines and airfoils) are also not covered.

The book's content not only provides a general description of fluid dynamics, but also teaches how to apply universal principles to build a mathematical model of a particular problem. The distinctive feature of M. Belevich's book is a somewhat non-standard approach of describing the dynamics of fluid from the point of view of the observer (chapter 15). It allows to underline some physical aspects of fluid mechanics which are usually not explicitly established in most textbooks.

The book is complemented by a carefully selected set of exercises. It provides consistent and self-sustained introduction to fluid dynamics, giving enough details to be used either in class or for self-study. It can be used to acquire knowledge in particular aspects of hydromechanics, and also as a source of inspiration for students, researchers and teachers in the field of classical fluid mechanics.

Ilya Rivin
Environmental Modeling Center
National Weather Service
National Oceanic and Atmospheric Administration
USA

PREFACE

This book presents the basis of the classical fluid mechanics and its content corresponds to a one-semester course which I am teaching from past several years to the 2nd year students of the Russian State Hydrometeorological University in St. Petersburg.

The goal of this book is twofold. Firstly, I wanted to provide a reader with a holistic idea of the fluid model and the way it is constructed. To show him, how the model of the fluid is developed, what main hypotheses lie in its basis and what general conclusions based on observations (the so-called laws of nature) make up the model. Secondly, I wished to demonstrate some possible modifications of the initial model which either make the model applicable in some special cases (viscous or turbulent fluid) or simplify it in accordance with peculiarity of a particular problem (hydrostatics, two-dimensional flows, boundary layers, *etc.*).

The whole theoretical material of the book naturally falls into two parts. The first part is fully dedicated to development of the model of the fluid in the Cauchy form. Here, the basic notions are introduced, main hypotheses are discussed and necessary postulates, which actually make up the model of continuum, are formulated. Non-coordinate tensor form of equations is actively used. This shortens formulas and makes results more readable. With that end in view, a brief introduction in tensor analysis is given in Ch.4. This part results in derivation of the perfect fluid model which turns out to be the simplest although quite efficient model.

In the second part of the book the most important modifications of the developed model are considered. First of all this concerns the redefinition of the stress tensor which is needed when viscosity is taken into account. Another important modification is connected with averaging of equations of the model which is necessary in case of turbulent flows. The concept of the boundary layer is also rather fruitful. Both laminar and turbulent boundary layers are discussed in Ch.14.

It is clear, that all this does not exhaust theoretical fluid mechanics, and that in the study of many important problems, it is necessary to refer to other books, at times rather special. However the basis of all such particular cases of the fluid mechanics is the same, and this book is aimed to discuss this topic.

Exercises and problems which are solved by students in practical classes are integral part of this book. They are chosen so as to teach students to work with complex systems of differential equations, since different fluid models are just such. We are training skills in writing equations in vector-matrix form, transition to component form of notation, applying of the index summation convention. Special attention is paid to formalizing of a verbal description of a problem (choice of coordinate systems, their orientation, accounting directions and symmetries inherent in the problem, *etc.*) as well as the mathematical problem posing. The system of fluid mechanics equations is quite complex and does not have analytic solutions in most interesting cases. Therefore, the main goal of these exercises is to elaborate the ability to see a particular problem from different viewpoints and to estimate its possible simplifications.

Theoretical fluid mechanics is very mathematized discipline, so the reader must meet certain requirements. Knowledge of the following topics of algebra and calculus is assumed: determinants, matrices, eigenvalue problem, vector spaces, calculus, vector analysis, differential equations.

All required information about tensors is given in Ch.4.

CONFLICT OF INTEREST

The editor declares no conflict of interest, financial or otherwise.

ACKNOWLEDGEMENTS

Besides my explicit and implicit teachers, I would like to especially thank those of my friends and colleagues, who patiently and sometimes willingly discussed with me various aspects of this course. These are primarily Dr. S.A.Fokin and A.A.Tron'. Their remarks, comments and suggestions were very helpful.

M.Belevich
St.Petersburg
Russia

Part I
Model of Continuum

A scientific law is a statement ... possessing such attributes: 1) it is true only under certain conditions; 2) under these conditions, it is true always and everywhere without any exceptions...; 3) conditions under which this statement is true, are never realized in fact fully, but only partially and approximately. Therefore, it's impossible to say literally that scientific laws are found in the study of reality (are discovered). They are devised (are invented) based on the study of experimental data in such a way that they then may be used to obtain new judgments of given judgments on the reality (including, for prediction purely logical way). Scientific laws themselves can not be verified and can not be refuted by experience. They can be justified or not, depending on how well or poorly they perform the above-mentioned role.

A. Zinoviev
Yawning Heights

CHAPTER 1

Bodies and Their Characteristics

Abstract: This introductory chapter tries to explain what we are going to do, what do notions such as fluid and a model of a physical phenomenon mean, what for, such models are developed, and what features of a phenomenon a model should be necessarily able to describe and so on.

Keywords: Body, Body configuration, Coordinate basis, Event, Fluid, Force, Frame of reference, Mass, Model, Motion, Place, Space-time continuum, System of coordinates, Trajectory, World-line.

1.1. INTRODUCTION

All natural-science disciplines have one and the same way of development: from gathering of facts, their classification, towards formalization of an object of research and its modeling. Fluid mechanics has attained a considerable progress in this regard. A lot of information has been collected over past centuries about properties and behavior of those objects which we now call gases, liquids and solids. Common features and differences of objects of study have been revealed as a result of systematization of this data. Basic properties of studied phenomena, and laws which these properties obey have been understood and formulated. Mathematical models which allow one to describe observations and predict changes of the form, position in the space and some other properties of simulated objects[1] have been constructed. These models essentially are the subject of the present course.

Our task is to examine hypotheses which define objects of study, axioms[2], underlying the described models, and the consequences arising out of these hypotheses and axioms. We emphasize again: the subject of the course are mathematical models of fluid dynamics. It is necessary to understand, that studying models of fluids (as well as models of any other object or phenomenon) is by no means the same thing, that the studying of real fluids themselves. The latter is an area of interests of experimenters and naturalists. In contrast, the work of a theorist consists in modeling of natural phenomena based on study of observational and experimental data.

Any model contains only what was laid by its creator. As a result, its connection with the initial natural phenomenon terminates, and each new observation may either correspond to it (*i.e.*, be described by this model) or not. At that, a contradiction with observation often does not indicate that the model is unambiguously bad. Probably the border of its region of applicability has been reached. This means that we have reached the border of the range of variation of parameters within which the hypotheses and axioms underlying the model, make sense. If the region of applicability of the model is unsatisfactory, the model should be modified or, at least, a new one should be built. Fortunately, the science is quite conservative and emergence of truly new models happens rather rarely. So, the age of the most widely used hydrodynamic models is about two hundred years.

Any model is an attempt to give a simple description of a complex phenomenon. If so, then something in this description should be neglected. What characteristics have to be described and what may be omitted? What aspects of a phenomenon the model must necessarily take into account? The answer to this question is far from unambiguous, and the one, that can be considered as adopted for today is the result of centuries of selection of facts, their classification and comprehension. Ultimately one gets something like the following.

Usually it is regarded, that the most significant features of observed objects are their abilities:

⊠ to exist, *i.e.*, to occupy a place in the three-dimensional space of places;
⊠ to move, *i.e.*, to change their places over time;
⊠ to keep the state of uniform motion and prevent its change (this ability is usually described using the concept of mass);
⊠ to interact with other objects (to describe such interaction the concept of force is introduced).

Certainly, it is far not all properties of real fluids. However the nature is arranged such that, in order to know how and why a fluid flows it does not matter what color, say, or what odour it has. But the four properties indicated above appear to be decisive.

Thus, the fluid mechanics, as a branch of the theoretical physics, describes abstract concepts, which are called *bodies*, and are endowed with following traits:

1. *place* in the three-dimensional space of places;
2. *motion*, the ability to change place over time;
3. *mass*, the ability to keep the state of the uniform motion;
4. the ability to interact with other bodies *via forces*.

1.2. SPACE OF EVENTS AND FRAMES OF REFERENCE

We shall think of a body B as a set of points $\{X, Y, \ldots\}$. Any such point, say X, at each moment of time t occupies a place $P(t, X)$ in the three-dimensional space of places P_t.

A pair (time, place) $= (t, P(t, \cdot))$ is called an *event*[3], and the set of all such pairs is called the *space of events* W or the *space-time continuum* (or just the *space-time*). The space of events may be imagined as consisting of an infinite (uncountable) set of spaces of places $P(t, \cdot) \equiv P_t$ numbered by a real parameter t which we identify with *time*[4].

Over time a point X of the body occupies places $P_t(X)$ in corresponding spaces P_t. A totality of events $(t, P_t(X)) \in W$ associated with one and the same point of the body B makes up a curve $\lambda(t, X)$ which is called the world-line of this point of the body. Each point of the body has its own *world-line*[5].

If we assume that the body B always consists of the same points (*i.e.*, point of the body do not arise and do not disappear), then one and the same event cannot be associated with different points of the body, and therefore world-lines cannot intersect each other and/or merge. We will adhere to this point of view. Besides, the time is directed only from past to future, and therefore world-lines do not have points of self-intersection. The world-line of each point X of the body is a function of one variable t. We assume that this function is sufficiently smooth and is differentiable with respect to t at least twice.

Points of the body B occupy one or another totality of places $P_t(B)$ in spaces P_t depending on time t. The totality of world lines of the body points is called the *world-tube* $\lambda(t, B)$ of the body. An example of a world-line and a world-tube is shown in Fig. (**1.1**). Three-dimensional spaces of places are represented here in the form of pieces of planes for descriptive reasons. The space of events W is represented as an infinite set of three-dimensional (two-dimensional in the figure) spaces of places, "strung" on the time axis. Each time moment is associated with its unique space of places. In turn, a time moment is possible to regard as a number of corresponding space of places, each of which is a space of simultaneous events (all events associated with points of a certain space of places P_t occurred in one and the same time t).

What does it mean to describe a position of a body (or its point) in space of places or time? From the experience, it is known that an unambiguous indication of an event means determination of when and where it had happened, *i.e.* specification of date and place. Regarding events occurring on the Earth such specification, most often, means the indication of time elapsed since the birth of Christ, and also the direction and distance to any well-known point (to Mecca, for instance). Certainly, other variants are also possible.

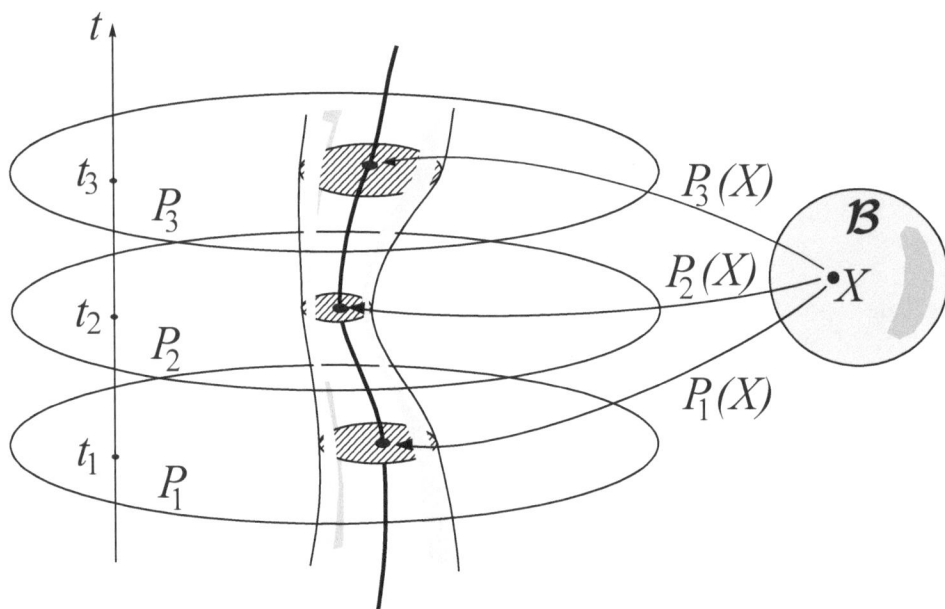

Fig. (1.1). The world-tube of the body $\mathcal{B}$ and the world-line of the point $X \in \mathcal{B}$. The spaces of places P_1, P_2, P_3 are spaces of simultaneous events. All events which belong to P_i occur at time t_i.

In any case, certain benchmark events are required, relative to which other events are determined. Ultimately, the events are considered to be described if an ordered set of numbers (say, (time, direction, distance)) is associated with each one, and benchmark events for which this set of numbers has specified meaning (for example, Christmas and some preselected point) are indicated. Such method of description of events is difficult to overestimate, as it opens the way to construction of quantitative (*i.e.*, mathematical) models of phenomena, allowing to replace the manipulations with events by operations with numbers, and this is a well-developed area of knowledge.

Nowadays, after centuries of doubts and disputes in classical science has prevailed a viewpoint, according to which the space of places is homogeneous and isotropic and time is homogeneous. In other words, the space-time continuum itself is considered to be devoid of benchmarks, relative to which it would be possible to calculate distances and directions[6]. In a situation when there are no absolute benchmarks, it is necessary to appoint them ourselves. People constantly use a large number of such benchmarks and choose one or another depending on a problem they face. Essentially, this means a choice of the zero time moment (The Creation of the World, New Year, beginning of a lesson, *etc.*) and the time unit, as well as a system of coordinates in the space of places (usually this is a polar system, but other systems are often used also; the origin of coordinates may be connected with the surface of the Earth, with the "fixed stars", with a microscope slide, *etc.*).

As a result, each event is unambiguously associated with an ordered set of four numbers[7] (time t and three coordinates of a place $\mathbf{x} = (x_1, x_2, x_3)$), which make sense of four coordinates in the space-time continuum $\mathcal{W}$ if the above-mentioned coordinate systems (in the time axis and in the space of places) are specified.

The ordered four of numbers $(t, \mathbf{x}) = (t, x_1, x_2, x_3)$ is an element of the four-dimensional real space $\mathbb{R}^4$. Mapping ϕ which assigns a point from $\mathbb{R}^4$ to an event from $\mathcal{W}$ is called *a frame of reference*

$$\phi: \mathcal{W} \to \mathbb{R}^4. \qquad\qquad (1.1)$$

The pair (*time, place*), *i.e.* an event, is associated with another pair (*a real number, an ordered triple of real numbers*). A totality of time moments is mapped on the real axis $\mathbb{R}^1$, the zero point of which determines the *origin* of the frame of reference. A real number t put in correspondence with a certain time instant, is called the *time* of this instant. Distance $|\, t_1 - t_0 \,|$ between two instants of time is called the time interval. If $t_1 > t_0$ then t_1 is called a later time than t_0.

An ordered triple of real numbers $\mathbf{x} = (x_1, x_2, x_3)$ associated with place $P_t(\cdot)$ unambiguously defines it and is called the *coordinates* of this place, in case the *coordinate system* in the space of places is specified. Thus, the system of reference, and, in general, any one-to-one mapping, is a rule that associates (one-to-one) elements of one set with elements of another one.

It should be understood that there is no certain element in the space $\mathbb{R}^4$ specially assigned for the event $(t, P_t(\cdot))$ from $\mathcal{W}$. A connection between $\mathcal{W}$ and $\mathbb{R}^4$ we make ourselves, choosing one or another frame of reference ϕ. Similarly, there is no special element in the space $\mathbb{R}^3$ rigidly associated with any point in the space of places. The choice of coordinate system gives us such an interrelation. But this is a choice! And the numbers x_1, x_2, x_3 will remain only numbers, if a certain coordinate system is not implied along with them. The quantity $\mathbf{x} = (x_1, x_2, x_3)$ sometimes is called a *radius-vector* of corresponding point of the space of places and the three numbers x_1, x_2, x_3 are regarded as its components. At that, all radius-vectors usually start at the origin of the selected coordinate system (if this system is Cartesian), and indicate those points, which they describe.

Often there is a temptation to merge together the space of places P and the three-dimensional space of real numbers $\mathbb{R}^3$, to think of them as a single whole, to identify them. As a rule such identification is very convenient. Even drawings become clearer if, among other things, they display a coordinate system (usually Cartesian). It is not surprising: after all, coordinates were invented for convenience. However, we should not forget that the space of places doesn't imply any coordinates, and an image of the space of places P in $\mathbb{R}^3$ strongly depends on the mapping $P \rightarrow \mathbb{R}^3$. For example, if the mapping is defined by a Cartesian coordinate system, the image of P coincides with the whole $\mathbb{R}^3$, but in case of, say, spherical coordinates the image of P will be already an infinite parallelepiped (Fig. **1.2**).

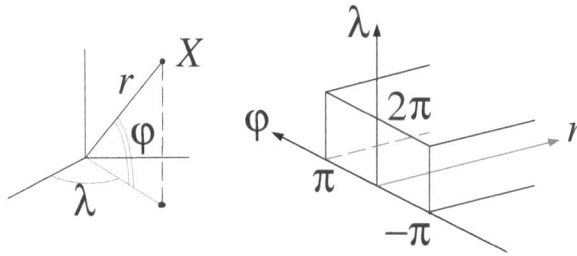

Fig. (1.2). Mapping of a space of places P (on the left) on $\mathbb{R}^3$ (on the right) in case of spherical coordinates. A semi-infinite parallelepiped in $\mathbb{R}^3$ is an image of P.

Exercise. Show that it is really so.

All of the above essentially means that the structure of a vector space is introduced on the set $\mathcal{W}$, *i.e.* the zero element (zero reference point) is chosen and rules of summation of elements and multiplication of elements by numbers satisfying a number of axioms are defined[8]. The same structure is induced on the sets of simultaneous events P_t where the zero element is the origin of the coordinate system, and all elements, by agreement, are called radius-vectors.

1.3. MOTION

Each point X of the body is associated with an infinite set of events, its world line $\lambda(t, X)$, a curve the points of which are numbered by a real parameter t, time. Time is changing, and events associated with the point of the body X at any given time are changing too. Tracking the change of events, we move over time along the world-line. The tangent vector $\vec{u} = d_t\lambda$ (Fig. **1.3**) defines the speed of this moving. The greater the speed of moving $|\vec{u}|$, the longer the tangent vector $d_t\lambda$.

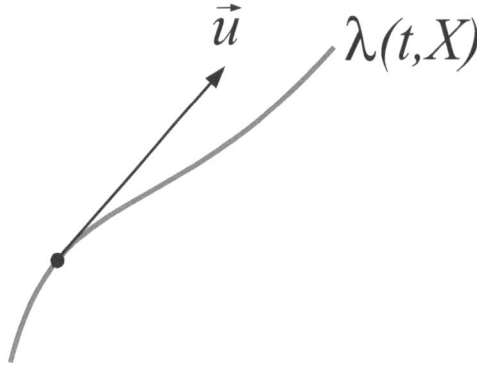

Fig. (1.3). Tangent vector $\vec{u}$ is the velocity of displacement along the world-line $\lambda(t,X)$.

Using the frame of reference ϕ, we endow all events with coordinates. Points of the world-line $\lambda(t,X)$ are getting coordinates $(t,\mathbf{x}(t))$ where $\mathbf{x}(t)=(x_1(t),x_2(t),x_3(t))$ are coordinates of the place, occupied at time t by the point X with respect to the chosen frame of reference . In this case, the tangent vector $\vec{u}$ may also be associated with a totality of numbers, its components

$$\vec{u} = d_t\lambda(t,X) \overset{\psi}{=} d_t(t,\mathbf{x}(t)) = d_t(t,x_1(t),x_2(t),x_3(t)). \tag{1.2}$$

Further we shall omit letters above equal signs indicating the selected frame of reference. It should be clearly understood that vectors do not depend on introduction of a frame of reference. They exist regardless to it. However, the components of vectors such, as they are defined here, arise only from the introduction of a frame of reference.

The vector $\vec{u}$ as any other derivative is a limit of the ratio

$$\vec{u} = \lim_{\Delta t \to 0} \frac{\lambda(t+\Delta t,X)-\lambda(t,X)}{\Delta t}.$$

After introduction of a frame of refrence the vector $\vec{u}$ may be written as

$$\vec{u} = \lim_{\Delta t \to 0} \frac{(t+\Delta t, \mathbf{x}(t+\Delta t)) - (t, \mathbf{x}(t))}{\Delta t}.$$

Here the numerator is a difference of elements of $\mathbb{R}^4$ and it is calculated componentwise. This gives

$$\begin{aligned}
\vec{u} &= \lim_{\Delta t \to 0} \left(\frac{(t+\Delta t)-t}{\Delta t}, \frac{\mathbf{x}(t+\Delta t)-\mathbf{x}(t)}{\Delta t} \right) \\
&= \left(1, \lim_{\Delta t \to 0} \frac{\mathbf{x}(t+\Delta t)-\mathbf{x}(t)}{\Delta t} \right) \\
&= \left(1, \lim_{\Delta t \to 0} \frac{x_1(t+\Delta t)-x_1(t)}{\Delta t}, \; \dots \right).
\end{aligned}$$

Using the notation $v_i = d_t x_i$ we have

$$\vec{u}(t, \mathbf{x}(t)) = (1, d_t \mathbf{x}(t)) = (1, v_1, v_2, v_3).$$

It remains to find out, with respect to what basis these numbers $(1, \{v_i\})$ have meaning of components of the vector $\vec{u}$. In other words, what are the vectors $\{\vec{e}_i\}_{i=0}^{3} = \{\vec{e}_0, \vec{e}_1, \vec{e}_2, \vec{e}_3\}$, a linear combination of which with weights $(1, v_1, v_2, v_3)$ unambiguously determines the velocity vector $\vec{u}$

$$\vec{u} = 1 \cdot \vec{e}_0 + v_1 \vec{e}_1 + v_2 \vec{e}_2 + v_3 \vec{e}_3? \tag{1.3}$$

Firstly, the basis vectors $\{\vec{e}_i\}$ and the vector $\vec{u}$ have to be elements of the same vector space (tangent, herein), in order the expression (1.3) makes sense. This means in particular that the point $(t, \mathbf{x}(t))$ of the world-line, is the zero vector of corresponding tangent vector space, and all vectors belonging to it, basic vectors including, begin at this very point. Secondly, the totality of vectors $\{\vec{e}_i\}$ should not contain the zero vector. Otherwise, the representation (1.3) is not unique and, therefore, $\{\vec{e}_i\}$ is not the basis. Thirdly, the components of the basis vectors with respect to themselves are

$$\vec{e}_0 = (1,0,0,0), \quad \vec{e}_1 = (0,1,0,0), \quad \vec{e}_2 = (0,0,1,0), \quad \vec{e}_3 = (0,0,0,1).$$

For example, it is easy to see that $\vec{e}_2 = (0,0,1,0) = 0 \cdot \vec{e}_0 + 0 \cdot \vec{e}_1 + 1 \cdot \vec{e}_2 + 0 \cdot \vec{e}_3$.

Now, we shall try to understand how the basis vectors $\{\vec{e}_i\}$ are located. The component $v_i = d_t \, x_i$ is the rate of change of i-th coordinates of the place when moving along the world-line. If $v_i = 0$ then this change is absent. Thus, if over time the point X of the body occupies places with unchanging coordinates its world-line λ is parallel to the time axis. The tangent vector $d_t \, \lambda = \vec{u}$ is also parallel to the axis t, it has components $(1,0,\,0,0)$ and, therefore, $\vec{u} = 1 \cdot \vec{e}_0$. But $\vec{u}$ is directed along the axis t. Hence, the basis vector $\vec{e}_0$ has the same direction (see, Fig. **1.4**).

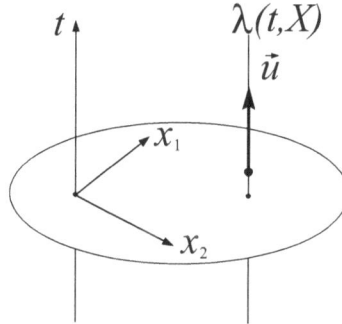

Fig. (1.4). The world-line of a point at rest relative to the coordinate system.

To find out the location of remaining basis vectors we shall study the projection of the world-line onto the space of places P. Projecting sets up a correspondence between the point of the world-line $\lambda(t, \, X) = (t, \mathbf{x}(t))$ and a point of the *parameterized curve*[9] $\mathbf{x}(t)$ in the space of places. The mapping χ: $(t, X) \mapsto \mathbf{x}(t)$ or

$$\mathbf{x}(t) = \chi(t, X) \tag{1.4}$$

is called the *motion* of the point X and the curve $\chi(t, X)$ is the *trajectory* of the point X in the space of places. At each time t the points of the body correspond to a totality of places $\chi(t, \mathcal{B})$ which is called *a configuration of the body* at this time. Thus, a configuration is a cross-section of the world-tube of the body at the given moment of time. Anyone, who has ever watched a running droplet of mercury or a spreading water puddle, easily will understand, what is a configuration of a fluid body. Bodies in both examplers may be considered all the time the same, but their configuration are constantly changing .

Since we have already decided not to consider intersecting world-lines, the mapping (1.4) is reversible, *i.e.*, there exists a map $\chi^{-1}:(t, \mathbf{x}) \mapsto X$. In other words, for any given time t and coordinates of the place $\mathbf{x}$, it is always possible to indicate a point X of the body

$$X = \chi^{-1}(t, \mathbf{x}), \tag{1.5}$$

which occupied this place at time t.

Vector $\vec{v}$ tangent to the curve $\chi(t, X)$ is the projection of the vector $\vec{u}$ on the same space of places. Its components with respect to the basis $\{\vec{e}_1, \vec{e}_2, \vec{e}_3\}$ are numbers (v_1, v_2, v_3), *i.e.*

$$\vec{v} = d_t\mathbf{x} = d_t x_1 \vec{e}_1 + d_t x_2 \vec{e}_2 + d_t x_3 \vec{e}_3$$
$$= v_1 \vec{e}_1 + v_2 \vec{e}_2 + v_3 \vec{e}_3.$$

For any basis vector, say $\vec{e}_3$, one finds that $\vec{e}_3 = 0 \cdot \vec{e}_1 + 0 \cdot \vec{e}_2 + 1 \cdot \vec{e}_3$ where $d_t x_1 = d_t x_2 = 0$ and $d_t x_3 = 1$. This means that $\vec{e}_3$ is the unit vector tangent to the coordinate curve[10] x_3. The remaining basis vectors $\vec{e}_1$ and $\vec{e}_2$ are unit vectors, tangent to coordinate curves x_1 and x_2 respectively (see. Fig. **1.5**). Basis composed of vectors tangent to coordinate curves is called *the coordinate basis*.

Finally, we obtain the following picture. Each point of the world line is connected with a *tangent vector space*, whose representative is a vector, tangent to the world line at this point. If the basis in such a vector space is constructed of the unit vectors, tangent to the coordinate curves (see, Fig. **1.6**), then the components of the velocity vector $\vec{u}$ (the rate of change of displacement along the world-line) will be equal to the time derivatives of corresponding coordinates of the point $\lambda(t, X) = (t, \mathbf{x}(t))$. At that, the last three components of $\vec{u}$ are the components of the vector $\vec{v}$, tangent to the projection of the world-line onto the space of places, *i.e.*, to the trajectory of the motion. Certainly, it is possible to change the basis, but then the components of both vectors $\vec{u}$ and $\vec{v}$ will change too.

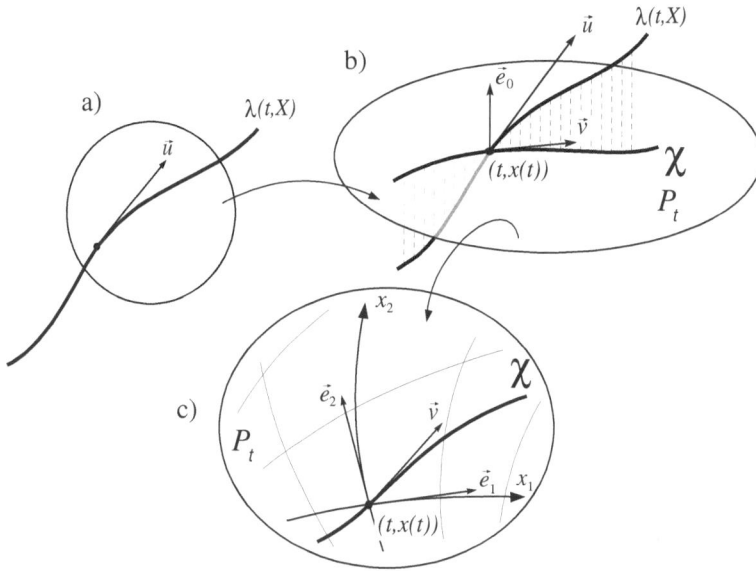

Fig. (1.5). a) The world-line of the point X of the body and the tangent vector $\vec{u}$; **b)** the trajectory of the motion of the point X is a projection of the same world-line λ onto the space of places P_t; the vector $\vec{v}$ tangent to the trajectory is a projection of the vector $\vec{u}$ onto P_t: $\vec{u} = (1, \vec{v}) = 1 \cdot \vec{e}_0 + \vec{v}$; **c)** coordinates in the space of places P_t and the coordinate basis $(\vec{e}_1, \vec{e}_2, \vec{e}_3)$, with respect to which $\vec{v} = (v_1, v_2, v_3)$.

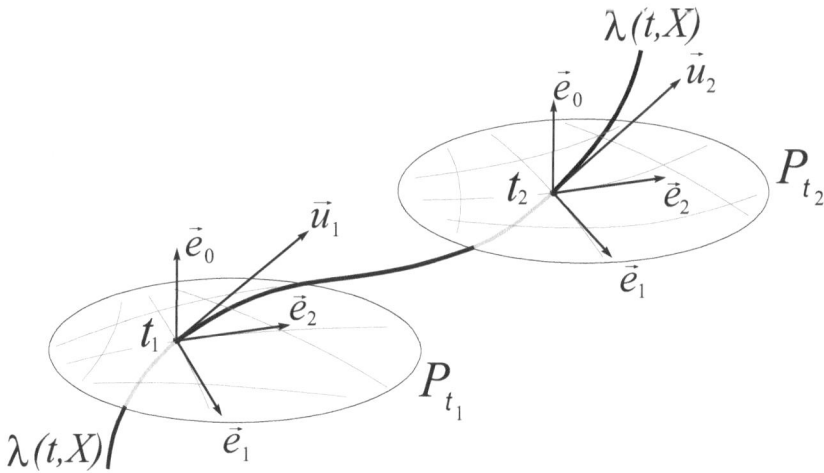

Fig. (1.6). Bases of tangent spaces at different points of the world line, which correspond to the time instants t_1 and t_2; vectors $\vec{u}_1$ and $\vec{u}_2$ are velocity vectors of a point X of the body in the space-time continuum.

1.4. MASS OF A BODY

Numerous observations show that all bodies, which are of interest to fluid mechanics, somehow or other have the ability to preserve the state of uniform motion and hamper its changing. This ability of a body is called its mass. It is so conservative characteristic of a body that it was noticed and formed the basis of all classical physics under the name of the *mass conservation law*.

The simplest way to describe any property of a body is to ascribe to a body some numerical value characterizing this property. Exactly so it is possible to do with the mass. Each body $\mathcal{B}$ is attributed to a real number $M(\mathcal{B})$ which is called the mass of the body. Thus, we automatically satisfy the mass conservation law, since the number, characterizing mass, is attributed to a body, regardless of time.

What numbers should be used to characterize mass? Since all bodies only hamper changes in the state of motion and never contribute to this, masses of all bodies must be numbers of the same sign. Traditionally the plus sign is chosen. Besides, since none of the bodies can prevent a change in the state of motion, the mass of a body is considered to be a finite number. Finally, another observed property of mass is associated with the so-called *separated bodies* (such naming is used to bodies, say $\mathcal{B}$ and $\mathcal{C}$ which have no common points: their intersection is empty, $\mathcal{B} \cap \mathcal{C} = \emptyset$). Total mass of any two separated bodies $\mathcal{B}$ and $\mathcal{C}$ is equal to the sum of both masses

$$M(\mathcal{B} \cup \mathcal{C}) = M(\mathcal{B}) + M(\mathcal{C}). \tag{1.6}$$

This property is called *additivity*.

Now let's formalize all the above-said. Let the set of all bodies with mass is denoted by Ω. This means that if any two bodies $\mathcal{B}$ and $\mathcal{C}$ are elements of Ω then $\mathcal{B} \cup \mathcal{C}$ and $\mathcal{B} \cap \mathcal{C}$ also belong to Ω. Since the intersection of two bodies may be empty, we assume that the empty set is also an element of Ω and its mass (and only its) is equal to zero. Now we define a bounded, real, non-negative, additive function M on Ω

$$M: \Omega \to \mathbb{R}^{0+}, \tag{1.7}$$

which we call the mass function. Here $\mathbb{R}^{0+}$ denotes the set of non-negative real numbers. The value of the mass function M on some element $\mathcal{B}$ of the set Ω is called the *mass $M(\mathcal{B})$* of the body $\mathcal{B}$.

However, the given definition does not specify the mass function M unambiguously. So, if M is a certain mass function and $c = \text{const} > 0$, then another function $c \cdot M$ is also a mass function, since it satisfies the definition (*i.e.*, it is also a finite, positive and additive function). Obviously, it is possible to introduce different mass functions on Ω, choosing this or that value of the constant c. Numerical values of masses of bodies will be different, but their ratios, at that, will not change:

$$\frac{M(\mathcal{B}_1)}{M(\mathcal{B}_2)} = \frac{\not{c}\, M(\mathcal{B}_1)}{\not{c}\, M(\mathcal{B}_2)}.$$

Generally speaking, we are interested not in the value of mass $M(\mathcal{B})$ which, as we have just seen, may be an arbitrary number, but in the ratio of $M(\mathcal{B})$ to the mass of a standard body $\mathcal{E}$. In particular, weighting, *i.e.* measuring of the weight W of the body, is based on the ratio of masses of a body $\mathcal{B}$ and a standard (weight) $\mathcal{E}$.

Indeed, due to the fact that $W(\mathcal{B}) = M(\mathcal{B})\, g$, $g = \text{const} > 0$, one has

$$\frac{W(\mathcal{B})}{W(\mathcal{E})} = \frac{M(\mathcal{B})\not{g}}{M(\mathcal{E})\not{g}}.$$

Thus, it is impossible to distinguish between two, proportional to each other mass functions by weighting. The choice of a particular mass function, in essence, is the choice of a standard, the mass of which is compared to the mass of another body. Recollect that, speaking about weight of a body, we call not only a number but also a unit of measure, *i.e.*, we explicitly specify the standard, which is assigned the mass, equal to unit.

1.5. FORCE

By this time, the observed inhomogeneous space of events we have agreed to interpret in the following way:

- the space-time continuum is homogeneous,
- inhomogeneities are identified with bodies,
- bodies tend to move uniformly.

It remains to explain why they do not move this way, *i.e.*, why they are moving non-uniformly, or, in other words, with acceleration?

Currently, the changing nature of a body motion is usually explained by its *interaction* with other bodies, and the mechanism of this interaction is described using the concept of force. A system of forces is introduced on the set Ω, and this means that each pair $(\mathcal{B}, \mathcal{C})$ of elements of Ω is associated with a quantity $\mathbf{f}_{\mathcal{B}}(\mathcal{C}, t)$ called the *force*, with which a body $\mathcal{C}$ acts on a body $\mathcal{B}$ at time t. In turn, the quantity $\mathbf{f}_{\mathcal{C}}(\mathcal{B}, t)$ will be the force with which the body $\mathcal{B}$ acts on the body $\mathcal{C}$.

The numerous observations of mechanical interactions have revealed a number of properties which the forces of interaction of separated bodies possess. For example, the following statements are valid:

1. The force of action $\mathbf{f}_{\mathcal{B}}(\mathcal{C}, t)$ is equal in magnitude to the force of action $\mathbf{f}_{\mathcal{C}}(\mathcal{B}, t)$ but is directed into the opposite side;
2. If the body $\mathcal{B}$ is affected by two bodies $\mathcal{C}$ and $\mathcal{D}$ then each of the forces $\mathbf{f}_{\mathcal{B}}(\mathcal{C}, t)$ and $\mathbf{f}_{\mathcal{B}}(\mathcal{D}, t)$ differs from the force $\mathbf{f}_{\mathcal{B}}(\mathcal{C} \cup \mathcal{D}, t)$.
   Finally, it is possible to assume that
3. Interaction of a body with itself is equivalent to the absence of interaction.

In case of interaction of a body $\mathcal{B}$ with a whole totality of bodies $\{\mathcal{C}_i\}$ it is often convenient to consider the interaction of $\mathcal{B}$ with one body, which is defined as the union of all bodies of the totality $\{\mathcal{C}_i\}$, *i.e.*, it is convenient to introduce the so-called *resultant* force $\mathbf{f}_{\mathcal{B}}(\cup_i \mathcal{C}_i, t)$. The resultant force may be determined for each body $\mathcal{B}$ of Ω using the notion of the *exterior of a body*[11], which means a body $\mathcal{B}^e$ such that $\mathcal{B} \cap \mathcal{B}^e = \emptyset$ and $\mathcal{B} \cup \mathcal{B}^e = \Omega$ *i.e.* a body from Ω whose points are not included in $\mathcal{B}$. Then the resultant force acting on $\mathcal{B}$ will be called the force $\mathbf{f}_{\mathcal{B}}(\mathcal{B}^e, t)$ which will be denoted $\mathbf{f}_{\mathcal{B}}(t)$.

All these properties of interactions lead to conclusion that for their correct description it is necessary to be able to sum them (*i.e.* to describe the combined action of two or more bodies on the third one) and multiply them by numbers (*e.g.*, the force $\mathbf{f}_{\mathcal{C}}(\mathcal{B}, t)$ from the property 1 may be interpreted as the force $\mathbf{f}_{\mathcal{B}}(\mathcal{C}, t)$ multiplied by -1). In other words, we need to come up with such rules of addition of forces and multiplication them by numbers, to be able to describe the action of two separated bodies $\mathcal{A}$ and $\mathcal{C}$ on the third body $\mathcal{B}$ as the sum of corresponding forces:

$$\mathbf{f}_{\mathcal{B}}(\mathcal{A} \cup \mathcal{C}, t) = \mathbf{f}_{\mathcal{B}}(\mathcal{A}, t) + \mathbf{f}_{\mathcal{B}}(\mathcal{C}, t), \qquad (1.8)$$

and the reaction force (see. property 1 above) as the product of force of action by the number (-1):

$$\mathbf{f}_{\mathcal{C}}(\mathcal{B}, t) = (-1)\, \mathbf{f}_{\mathcal{B}}(\mathcal{C}, t).$$

In mathematical arsenal there exists a developed tool for working with such objects. It is called the *vector space*. So if we want to correctly define the addition of forces and their multiplication by numbers, it would be convenient to endow the system of forces by the structure of the vector space[12]. The introduction of such a structure on the set Ω means the following.

- An arbitrary element $\mathcal{B} \in \Omega$ is selected and is associated with the zero vector 0, which is interpreted as the force $\mathbf{f}_\mathcal{B}(\mathcal{B}, t) = 0$ (see the above-mentioned property 3).
- All other elements, *e.g.*, $\mathcal{C}_i$ are associated with vectors, which are interpreted as forces of action $\mathbf{f}_\mathcal{B}(\mathcal{C}_i, t)$ of bodies $\mathcal{C}_i$ on the body $\mathcal{B}$.
- A rule is specified, according to which every two vectors $\mathbf{f}_\mathcal{B}(\mathcal{C}_1, t)$ and $\mathbf{f}_\mathcal{B}(\mathcal{C}_2, t)$ are associated with the third vector $\mathbf{f}_\mathcal{B}(\mathcal{C}_3, t) = \mathbf{f}_\mathcal{B}(\mathcal{C}_1, t) + \mathbf{f}_\mathcal{B}(\mathcal{C}_2, t)$, called their sum.
- A rule is specified, according to which each vector $\mathbf{f}_\mathcal{B}(\mathcal{C}, t)$ and a number α are associated with another vector $\mathbf{f}_\mathcal{B}(\mathcal{A}, t) = \alpha\, \mathbf{f}_\mathcal{B}(\mathcal{C}, t)$, called their product.
- These rules must satisfy the following axioms (see any textbook on linear algebra for details).

1. The zero vector 0 is such, that

$$\mathbf{f}_\mathcal{B}(\mathcal{C}, t) + 0 = \mathbf{f}_\mathcal{B}(\mathcal{C}, t).$$

2. Each vector $\mathbf{f}_\mathcal{B}(\mathcal{C}, t)$ has an *inverse vector* ($\mathbf{f}_\mathcal{B}(\mathcal{C}, t)$) such, that

$$\mathbf{f}_\mathcal{B}(\mathcal{C}, t) + (-\mathbf{f}_\mathcal{B}(\mathcal{C}, t)) = 0.$$

3. Multiplication by number 1 is the identity operation

$$1 \cdot \mathbf{f}_\mathcal{B}(\mathcal{C}, t) = \mathbf{f}_\mathcal{B}(\mathcal{C}, t).$$

4. *Commutativity* of summation

$$\mathbf{f}_\mathcal{B}(\mathcal{A}, t) + \mathbf{f}_\mathcal{B}(\mathcal{C}, t) = \mathbf{f}_\mathcal{B}(\mathcal{C}, t) + \mathbf{f}_\mathcal{B}(\mathcal{A}, t).$$

5. *Associativity* of summation

$$(\mathbf{f}_\mathcal{B}(\mathcal{A}, t) + \mathbf{f}_\mathcal{B}(\mathcal{C}, t)) + \mathbf{f}_\mathcal{B}(\mathcal{D}, t) = \mathbf{f}_\mathcal{B}(\mathcal{A}, t) + (\mathbf{f}_\mathcal{B}(\mathcal{C}, t) + \mathbf{f}_\mathcal{B}(\mathcal{D}, t)).$$

6. *Associativity* of multiplication

$$(\alpha\beta)\mathbf{f}_\mathcal{B}(\mathcal{C}, t) = \alpha(\beta\mathbf{f}_\mathcal{B}(\mathcal{C}, t)).$$

7. *Distributivity* of summation

$$\alpha(\mathbf{f}_\mathcal{B}(\mathcal{A}, t) + \mathbf{f}_\mathcal{B}(\mathcal{C}, t)) = \alpha\mathbf{f}_\mathcal{B}(\mathcal{A}, t) + \alpha\mathbf{f}_\mathcal{B}(\mathcal{C}, t)$$

8. *Distributivity* of multiplication

$$(\alpha + \beta)\mathbf{f}_{\mathcal{B}}(\mathcal{C}, t) = \alpha\mathbf{f}_{\mathcal{B}}(\mathcal{C}, t) + \beta\mathbf{f}_{\mathcal{B}}(\mathcal{C}, t)$$

After removing parentheses in the second term of the left-hand side of the axiom 2, this expression is usually written as a difference of vectors

$$\mathbf{f}_{\mathcal{B}}(\mathcal{C}, t) + (-1) \cdot \mathbf{f}_{\mathcal{B}}(\mathcal{C}, t) = \mathbf{f}_{\mathcal{B}}(\mathcal{C}, t) - \mathbf{f}_{\mathcal{B}}(\mathcal{C}, t) = 0.$$

Thus, the subtraction of vectors is defined *via* summation and multiplication by -1.

Actually all these axioms are quite natural. They make working with such new objects as vectors, simple and easy. If summation and multiplication satisfy these axioms, then it is possible to rearrange terms in sums (axiom 4), not to indicate the order of summation and multiplication (axioms 5 and 6), to remove parentheses or to factor out common factors (axioms 7 and 8) and so on.

Why does the body $\mathcal{B}$ appear everywhere? Only because, being interested in the influence of different bodies from Ω on the body $\mathcal{B}$, we have associated with it the zero vector. From the mathematical standpoint it is possible to take any other body instead of $\mathcal{B}$. At that, another vector space on the same set Ω will be obtained. But it must be kept in mind that addition is defined only for vectors belonging to the same vector space. If, say, two vector spaces are defined on the set Ω, then the sum of two vectors from these two different spaces is undefined, *i.e.*, is simply meaningless[13].

After introduction, of the structure of the vector space on the set Ω the rules of vector addition and multiplication by numbers are at our disposal. The interpretation of these rules (*i.e.*, what particular vector should be regarded as a sum of two others), is defined by the problem[14] to be solved. Since a vector is interpreted within the present case, as a force, then, according to observations, the sum of two vectors is a force, which coincides with a diagonal of a parallelogram constructed on both vectors. Another problem may require another interpretation.

NOTES

[1] In fluid mechanics it is accepted to call fluids all the objects under study, regardless of how they are perceived by our senses.

[2] Since we deal with mathematical models of physical phenomena, these axioms should correspond to the so-called laws of nature, extracted from observational data.

[3] Hereafter the icon (·) is a "place holder". It should be replaced by an object out of previously specified set.

[4] As a good illustration for the space of events a film strip with any filmed episode may be used. Let's assume that each frame fixes simultaneous events in different points of the space of places. But there are a lot of frames! What are depicted on them? One and the same space of places, where there is time, and everything lives, everything changes? Or we see images of different spaces of places, each of which is associated with its own moment of time and not with any other, or, in other words, in which there is no time (where the time itself is only a chain of successive spaces of places)? There is no definite answer to this question. Any viewpoint has its right to exist. The first one is more traditional, while the second, which came from the theory of relativity, looks more universal and therefore seems to be more promising.

[5] To illustrate this idea, take a recorder chart. One-dimensional motion of the pen turns into a flat curve in the space of places (chart). Each point of this curve (*i.e.*, a position of the pen, is associated with a particular time) is a projection of the current event onto the space of places. World-lines contain all the information about the movement. An example of this is a gramophone record that stores a copy of a segment of the projection of the world-line of a tip of the needle of sound recording apparatus. Those sounds that forced the needle to move, may be extracted again from the information stored in the sound groove of the record.

[6] From this viewpoint, the heterogeneity and anisotropy, observed everywhere, are associated with objects, placed in the space, which is deprived of any inhomogeneity. As for time, it is assumed to be homogeneous, since (and so long as) there is no reason to consider it otherwise.

[7] The ordering means that not only values of numbers are important, but also an order in which they are arranged. So, in the ordering we have adopted, the first number is always a time value. Two similar sets of unequal numbers, differing in their order only are considered to be different.

[8] It is supposed that the reader is "in the know" what is a vector space, its basis, vector components with respect to basis and so forth. Briefly the definition of a vector space is given below on p.15 as applied to forces.

[9] A parameterized curve in the space of places P will be called a differentiable mapping of an open subset of $\mathbb{R}^1$ into P. In fact this means that parameterization of a curve is a sequential numbering of all its points by real numbers (values of parameter). Each world-line itself is a parameterized curve. The time t is the parameter which labels (numbers) points of the curve.

[10] Keep in mind that coordinate lines are a special case of coordinate curves, all points of which differ in values of one coordinate only.

[11] This concept and term are borrowed from the book [1].

[12] The physical interpretation of such a structure, used herein, is not unique. The possibility of other interpretations explains the widespread use of the concept of the abstract vector space.

[13] Later on, when such sums also will be needed, they will be specially defined (see, p.50). At present we don't consider them.

[14] From the mathematical viewpoint it is indifferent what are the above two rules, as long as they satisfy the given axioms. However, since this mathematical construction is used herein to describe a certain physical reality, these rules should be chosen so as to allow the desired physical interpretation.

CHAPTER 2

Basic Hypotheses and Laws

Abstract: We try to show the necessity of accepting the fundamental continuity hypothesis, which implies introduction of such new notions as integral parameters of a medium and their densities. We also discuss a configuration of a body and its deformation as well as methods of its description. The concepts of the Lagrangian and Euler coordinates are considered.

Keywords: Causality principle, Conservation law, Continuity hypothesis, Continuum, Current configuration, Deformation, Euler coordinates, Integral parameter, Knudsen number, Lagrangian coordinates, Local parameter (density), Reference configuration.

2.1. THE CONTINUITY HYPOTHESIS

The description of motion of a body means indication of explicit dependence of position $\mathbf{x}(t) = \chi(t, X)$ on time, for all particles X which make up this body. Moreover, the size of these particles must be small enough for they could be associated with points. How many particles make up the body, and how many of the above-mentioned dependencies it is necessary to obtain?

Let $\mathcal{P}_1$ be a part of the body $\mathcal{B}$. This means that $\mathcal{P}_1$ preserves all physical properties of $\mathcal{B}$. Let $\mathcal{P}_2$ be a similar part of $\mathcal{P}_1$ *i.e.* $\mathcal{P}_2 \subset \mathcal{P}_1$. And so on. How long one can continue building a chain $\mathcal{B} \supset \mathcal{P}_1 \supset \mathcal{P}_2 \supset \ldots \supset P_k \supset \ldots$ such that each part $\mathcal{P}_k$ possesses physical properties of $\mathcal{B}$? Is there some indivisible part X of $\mathcal{B}$ or division may be continued infinitely?

The answer is known: yes, there is. However, even in a drop of water there are so many of such extremely small indivisible particles (they are molecules, according to modern view) that it is impossible to describe individually the movement of each of them. The required number of dependencies $\mathbf{x}(t)$ describing the motion of particles far exceeds human capabilities. Given this, we either have to restrict the consideration to a small number of particles, losing sight of the object itself (liquid or gas) or obtain unthinkably complicated challenge. The situation is aggravated by the fact that in nature, the particles of the body are not independent and demonstrate complex interaction with each other. It is not always possible to neglect this interaction.

Unexpected exit from the impasse is the assumption of infinite divisibility of a body, *i.e.*, the assumption that it is not discrete within its configuration and, besides, no matter how small a part of the body is, it retains all physical properties of initial body. The paradox is that the problem, which seemed to be beyond one's strength because of the huge number of particles, may be simplified if mentally we increase the number of particles within the body, and will think of it as consisting of an infinite or even uncountable set of points. This is explained by the fact that giving up the consideration of the body as a totality of a large but finite number of particles and replacing it with the so-called continuum, we obtain on the one hand the possibility to use powerful research tool (mathematical analysis, theory of differential equations, *etc.*), and on the other, we approximately take into consideration inter-molecular interactions. To make all this possible, it is necessary to accept the assumption, known as the continuity hypothesis.

We shall call a body, a *continuum* if it satisfies the *continuity hypothesis*, *i.e.*, if the following assertions are true:

1. The body is an *everywhere dense set of points*.
2. Properties of the body are described by *differentiable functions* of the points of the body.

After acceptance of this hypothesis, formerly separate and independent particles of a body (*e.g.*, molecules) become connected with each other, because all properties of the body now are described by differentiable functions (this is the assertion 2). Now, knowing the behavior of a function at any point, we have some idea what happens to it at points in a vicinity of the given point, since each such function may approximately be written in the form of a segment of the Taylor series[1]. In order to have this opportunity, we needed to substitute a finite totality of particles for an infinite dense set of points. Additional dummy points fill gaps between physical particles of the body. The first assertion just means that there are no isolated points in the body: in any vicinity of each point of the body there is at least one point of the body more (for exact wording, see courses of mathematical analysis, *e.g.*, [2]).

In what case the real fluid may be regarded as continuum? Is it possible to call continuum a tea in a glass, an air in a room, a cluster of stars in the sky? Apparently, an object may be replaced with a continuous medium when it is not very sparse. However, the sparseness is a relative concept: water or air near the surface of the Earth do not seem to be too sparse. But, after all, we can study the motion of, say, a dozen of molecules of the same water. The medium in this case remains the same, the distances between molecules do not change, but it seems unlikely that the means, suitable to describe the behavior of a dozen of molecules, are also suitable in case of uncountable set of points.

To all appearances, it is important not only, what particles we consider and on what distances from each other they are but also how many of them there are. The number of particles that fall within our field of view is defined by a phenomenon that we study, by its characteristic spatial scale. If we study motion of a dozen of water molecules, most likely it cannot be replaced by continuum but if we consider the motion of the ocean of water, then without large errors, this object may be regarded as continuum. It is clear that the continuity hypothesis, like any other, has its limits of applicability: not every object or not always may be regarded as continuum. The criterion here is the ratio of the average distance l between particles[2] to the spatial scale of the phenomenon under consideration L. This is the so-called, *Knudsen number*[3],

$$Kn = \frac{l}{L} \cdot$$

It characterizes relative sparseness of the studied substances. When $Kn < 0.1$ the size of the region occupied by the studied body is large compared to the distance between particles of substance, and the use of the continuity hypothesis is considered acceptable[4]. If $Kn > 1$ we deal with the so-called free molecular flow. This is a scope of the kinetic theory of gases. Finally, the range $Kn \in [0.1, 1]$ is a transitional region. Here spatial scales of the body only slightly exceed distances between particles.

2.2. INTEGRAL PARAMETERS AND THEIR DENSITIES

Acceptance of the continuity hypothesis and understanding of a body as a continuous medium means that we are interested mainly in macroscopic processes in which a large number of real particles of liquid is involved. It is supposed to describe these processes explicitly. Everything that occur at the micro-level (where now we have a lot of dummy points), either should not be taken into account, or should be described parametrically (*i.e.* should be characterized jointly by the value of some parameter). Such an approach requires introduction of a number of new concepts and, first of all, *integral parameters* and *local parameters* or *densities*.

2.2.1. Parts of a Body

A closed surface that separates points of a connected subset[5] of some body $\mathcal{B}$ from the rest of its points, will be called the *boundary surface*, and both subsets will be called *parts of the body*. Of course, each part of a body, in turn, may be regarded as a body. There is a theorem (see *e.g.*, [4]), according to which a boundary surface at any time consists of the same points of a body. A surface possessing such property is called *kinematic boundary*. Thus, boundary surface, as well as surfaces of body configurations, are kinematic boundaries.

2.2.2. Integral Parameters of Continuum

If a quantity (for the present we shall denote it by Π), is a characteristic of a body as a whole, it is called the integral parameter of the body. As an example of such a parameter one may consider the mass which characterizes the ability of a body to prevent changes in the state of rest or uniform motion (what is just the same).

Formally, the *integral parameter* Π is defined as a bounded, real-valued, additive function defined on the set Ω (see p. 12).

In this definition, unlike the one that was given to the mass, the requirement of non-negativity is absent, *i.e.*, alternating-sign parameters as well as the dependence on time are allowed. Beside mass, other integral parameters may be considered. The volume $V(t,\mathcal{B}) = V(\chi(t,\mathcal{B}))$ of a configuration $\chi(t,\mathcal{B})$ of a body $\mathcal{B}$ is an example of time-dependent positive integral parameter, and the electric charge is an example of alternating-sign parameter. Further we will need to consider some of such parameters.

Suppose some integral parameter Π is defined for a body $\mathcal{B}$. Then it is also defined for any of its part $\mathcal{P} \subset \mathcal{B}$. At the same time, due to the additivity of the integral parameter, the inequality $|\Pi(t, \mathcal{P})| < |\Pi(t, \mathcal{B})|$ holds. Consider an infinite sequence of body parts $\{\mathcal{P}_k\}_{k=1}^{\infty}$ such that each successive element of the sequence is a subset of the previous one, and they all have only one common point X, *i.e.*, $\mathcal{P}_{k+1} \subset \mathcal{P}_k \, \forall k$, and $\cap_k \mathcal{P}_k = X$. Absolute values of parameters $\Pi^k = \Pi(t, \mathcal{P}_k)$ form a monotonically decreasing numerical sequence $|\Pi^1| > |\Pi^2| > \ldots > |\Pi^k| > \ldots$ bounded below by zero. It is known (see, *e.g.*, [2]) that such sequence has a limit $\lim_{k \to \infty} |\Pi^k| = 0$, *i.e.*, the value of an integral parameter at a point of continuum is zero. The mass $M(X)$ of any point of a body, as well as the volume $V(\chi(t, X))$ occupied by this point, are equal to zero. For this reason, integral parameters cannot characterize the continuum at a point.

Meanwhile, the limit of the ratio of the integral parameter of a part $\mathcal{P}_k$ of the body to the volume $V^k = V(\chi(t, \mathcal{P}_k))$ occupied by this part

$$\lim_{k \to \infty} \frac{\Pi^k}{V^k} = \pi(t, \chi(t, X)) = \pi(t, \mathbf{x}(t)), \tag{2.1}$$

is generally differs from zero and is a function of the point. The value of this limit $\pi(t, \mathbf{x}(t))$ is called the *density* of the integral parameter Π at point $(t, \mathbf{x}(t))$. For example, the *mass density* ρ is by definition

$$\rho(t, \mathbf{x}(t)) = \lim_{k \to \infty} \frac{M(\mathcal{P}_k)}{V^k}. \tag{2.2}$$

Unlike the mass, the mass density is a characteristic of a point of the subset of the space of places occupied by the body configuration at some given time. Thus, it is a *local characteristic*.

If density π of a parameter in some configuration χ is known, it is possible to determine the integral parameter Π by calculating the integral over the points of this configuration:

$$\Pi(t, \mathcal{B}) = \int_\chi \pi(t, \mathbf{x}(t)) dV. \tag{2.3}$$

Here dV is the infinitesimal volume element of the configuration. For example, the mass of a body may be obtained according to this formula as a value of the integral:

$$M(\mathcal{B}) = \int_\chi \rho(t, \mathbf{x}(t)) dV. \tag{2.4}$$

Since, by definition, it does not depend on time, this argument is omitted here.

2.2.3. Integral Characteristics of the Spatial Domain

Points of a body $\mathcal{B}$ occupy places (in the space of places) in accordance with the current configuration $\chi(t, \mathcal{B})$. Since the mapping χ is assumed reversible the domain $\chi(t, \mathcal{B})$ at time t and the body $\mathcal{B}$ are associated one-to-one: we say "body $\mathcal{B}$ at time t" and imply the region $\chi(t, \mathcal{B})$ in the space of places; we say "domain $\chi(t, \mathcal{B})$ at time t" and imply the body $\mathcal{B}$ located there at this moment.

If an integral parameter $\Pi(t, \mathcal{B})$ is connected with a certain body, then at any given time t this parameter may be considered not only as a characteristic of the body, but also as a characteristic of the domain $\chi(t, \mathcal{B})$, occupied by this body in the space of places. Thus,

$$\Pi(t, \mathcal{B}) = \Pi\big(\chi(t, \mathcal{B})\big).$$

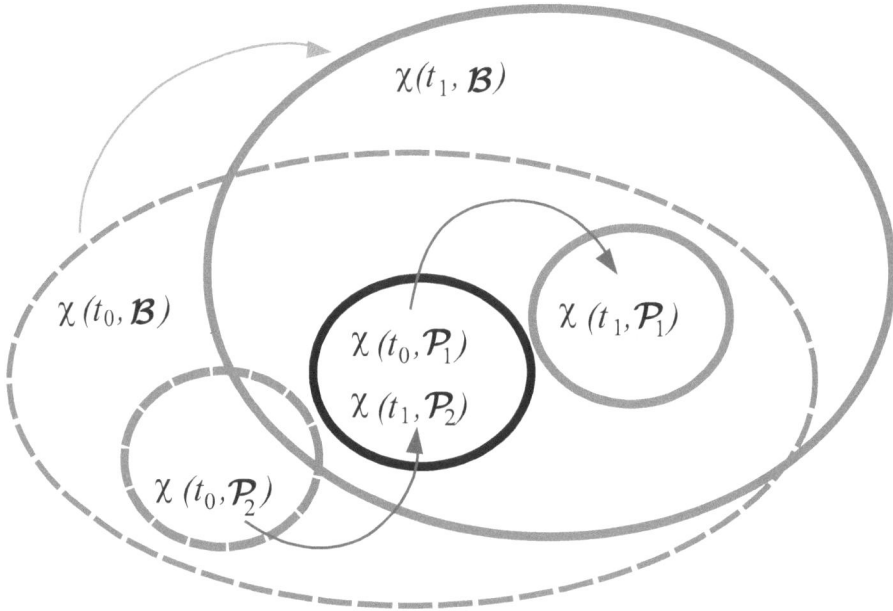

Fig. (2.1). At different times the space area bounded by the black line is the configuration of various parts of the body B. At time moments t_0 and t_1 this area is occupied by the parts P_1 and P_2, correspondingly.

Let $\mathcal{P}_1 \subset \mathcal{B}$ be a part of the body $\mathcal{B}$ and $\chi(t_0, \mathcal{P}_1)$ its configuration at time t_0. In general, at any other time t_1 this part of the body will occupy a new region $\chi(t_1, \mathcal{P}_1)$ and its former configuration (spatial domain) will be, in turn, occupied by another part of the body, for example, $\mathcal{P}_2$ *i.e.* $\chi(t_0, \mathcal{P}_1) = \chi(t_1, \mathcal{P}_2)$ (see Fig. **2.1**). Each spatial domain at different times is a configuration of various bodies or parts thereof. For this reason, when speaking about some parameter characterizing spatial domain at some time t, it is convenient to associate it not with the configuration of a part of the body (because here it does not matter which one), but with the volume V of this domain, *i.e.*, to consider the quantity $\Pi(t, V)$. In this case the integral parameter $\Pi(t, V)$ is associated with corresponding density by the relation

$$\Pi(t, V) = \int_V \pi dV. \tag{2.5}$$

2.2.4. About Differences

$\Pi(t, \mathcal{P})$ *vs* $\Pi(t, V)$.

Let us emphasize the difference between $\Pi(t, \mathcal{P})$ and $\Pi(t, V)$ once again.

$\Pi(t, \mathcal{P})$ is a characteristic of a body or its part, *i.e.*, a set of points, whose behavior we investigate regardless of the region where they are located;

$\Pi(t, V)$ is a characteristic of a spatial domain of a volume V, *i.e.*, a set of places, no matter what points of a body these places occupy.

If both parameters are considered as functions of time, then $\Pi(t, \mathcal{P})$ shows the change of Π in the body $\mathcal{P}$ over time (*e.g.*, change of the volume occupied by the configuration of the body). In turn, $\Pi(t, V)$ shows the evolution of the parameter Π in the volume V (for example, the change of the mass of a substance contained in a given volume).

Note that different meanings have integral parameters, only, but not corresponding densities. The function $\pi(t, \mathbf{x})$ in both formulas (2.3) and (2.5) is the same. Only in the first case, arguments of the density are coordinates $(t, \mathbf{x}(t))$ of a point of a world line, and in the second case, arguments are coordinates $(t, \mathbf{x})$ of a fixed point of the space-time. According to the accepted continuity hypothesis, the mass density like density of any other integral parameter, is considered to be a differentiable function of all its arguments, *i.e.*, the density varies from point to point (within configuration of a body) smoothly without jumps and breaks.

Integral Parameter vs Local Parameter

Also, it should be clearly understood the fundamental difference between integral and local parameters of the medium. The integral parameter is a measurable quantity. Moreover, we are able to measure only integral parameters. Corresponding mathematical notion is called the measure. The mass, volume, charge, *etc.*, are different measures of a body (medium). Correspondence here is unambiguous: if something is a measure then it may be measured, and if something is measurable, then it is a measure, *i.e.*, integral parameter of the medium, no matter how small volume it occupies (but not a point because a medium is always infinitely many points, by definition).

On the contrary, the density or local parameter is not measured ever, but is always calculated, already at least because the very point of the body in which it is determined theoretically in nature may not exist (real bodies are discrete at the micro level). After all, in order to avoid this nuisance, the continuity hypothesis was invented. Thus, the density is a theoretical, speculative design. Even if it is claimed that some device measures a density π, actually, it does not do it. It measures not the function of a point which we have defined here, but the ratio of two measures: the ratio of the integral parameter $\Pi(t, \mathcal{P}_k)$ of a small part of a body to the volume V^k occupied by this part. In other words, it measures some average density of the body $\mathcal{P}_k$. In case of homogeneous continuum it does not matter, which density to use, real or average. But if the continuum is heterogeneous? Inevitably there is a measurement error and, in general case, always non-zero.

2.3. DEFORMATION: THE EULER AND LAGRANGIAN COORDINATES

As it was already mentioned, the volume V of a configuration $\chi(t, \mathcal{B})$ is also an integral parameter but, unlike mass, its value is attributed not to the body, but to the totality of places, occupied by this body, *i.e.*, to the configuration of the body. For this reason, the volume of the configuration is, in general, a variable quantity. The density of this parameter, found using (2.3), is equal to unit everywhere, and therefore non-informative.

If, however, we select some configuration among a multitude of configurations corresponding to different times, then it becomes possible to study variation of the volume occupied by different configurations of the body with respect to the volume of the selected configuration. Such configuration is called the *reference configuration* κ. Most often the configuration, which corresponds to the initial moment of time t_0, is used:

$$\kappa(\mathcal{B}) = \chi(t_0, \mathcal{B}). \tag{2.6}$$

All other configurations are called *current configurations*.

Coordinates of places occupied by points of a body in the reference configuration, traditionally are called their *Lagrangian coordinates*[6] (*Lagrangian variables* or sometimes *Lagrangian labels*). We shall denote them by capital letters $\mathbf{X} = (X_1, X_2, X_3)$. Thus, for a point X of a body we have

$$\mathbf{X} = \kappa(X) = \chi(t_0, X). \tag{2.7}$$

The Lagrangian variables specify coordinates of points of a body in one fixed time moment and therefore, do not depend on time. Coordinates of the same points of a body in current configurations are called their *Euler coordinates (variables)*[7]. They are functions of time, since positions of points of a body at different times are generally different.

Given that the Euler coordinates of the same point X of the body at time t is $x(t) = \chi(t, X)$ and all configurations, by agreement, are reversible, we find the connection between the Lagrangian and Euler coordinates of the same point of the body. Namely, from (2.7) we find $X = \kappa^{-1}(\mathbf{X})$ and further,

$$\mathbf{x}(t) = \chi(t, \kappa^{-1}(\mathbf{X})) = \chi_\kappa(t, \mathbf{X}), \qquad \mathbf{x}(t_0) = \mathbf{X}, \tag{2.8}$$

i.e. the Lagrangian coordinates of the point X of the body are its Euler coordinates at the initial moment of time.

The time-dependent mapping $\chi_\kappa : (t, \mathbf{X}) \mapsto x(t) = \chi_\kappa(t, \mathbf{X})$ describes shifts of positions of points of the body when changing over from the reference configuration $\kappa(\mathcal{B})$ to the respective current one $\chi(t, \mathcal{B})$ or *deformation of the reference configuration*. It is called the *deformation* of the body $\mathcal{B}$ during time $(t - t_0)$ with respect to the reference configuration (Fig. **2.2**).

Please note that both mappings $\chi(t, X)$ and $\chi_\kappa(t, \mathbf{X})$ give current (Euler) coordinates of a point of the body. And if in the first case this point is explicitly specified, in the second it is given by its coordinates in the reference configuration (Lagrangian coordinates). Sometimes, instead of (2.8) we shall use the shorthand $\mathbf{x}(t) = x(t, \mathbf{X})$, which means that every trajectory $\mathbf{x}(t)$ is associated with the point of the body labeled with the Lagrangian coordinates $\mathbf{X}$.

Once the reference configuration is selected, it may be used to calculate the quantity characterizing locally current deformation of the reference configuration

$$J(t, \mathbf{x}(t)) = \lim_{k \to \infty} \frac{V(\chi_\kappa(t, \mathcal{P}_k))}{V_\kappa^k}. \tag{2.9}$$

Cf. this formula with (2.1) or (2.2). Here $V_\kappa^k = V(\kappa(\mathcal{P}_k))$ is the volume of $\mathcal{P}_k$ in the reference configuration. The geometric meaning of this quantity is the relative change of an infinitesimal volume at the change over from the reference configuration to the current one .

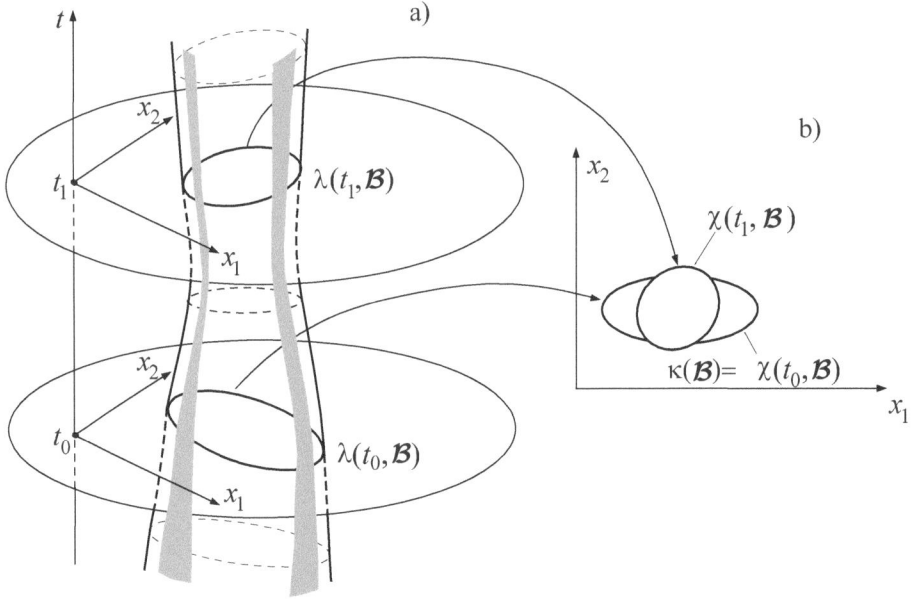

Fig. (2.2). a) Cross-sections of the world-tube of the body $\mathcal{B}$ corresponding to the reference time moment $t_0 : \lambda(t_0, \mathcal{B}) = (t_0, \kappa(\mathcal{B}))$ and some current time $t_1 : \lambda(t_1, \mathcal{B}) = (t_1, \chi(t_1, \mathcal{B}))$; **b)** deformation of the reference configuration $\kappa(\mathcal{B})$ into the current one $\chi(t, \mathcal{B})$. Both configurations are the world-tube cross-sections, projected onto some (and, hence, any) space of places.

The volume of the current configuration, by definition, is a quantity equal to

$$V\big(\chi_\kappa(t, \mathcal{B})\big) = \int_{\chi_\kappa} dV,$$

where the integration is performed over points of the current configuration. On the other hand, the comparison of the formulas (2.9) and (2.1) shows that the value of J may be regarded as the "density" of the volume of current configuration χ_κ with respect to the volume of the reference configuration κ. Then, using the formula (2.3) we compute

$$V(\chi_\kappa(t, \mathcal{B})) = \int_\kappa J \, dV.$$

Here the integration is performed over the points of the reference configuration. Comparing both formulas, we obtain

$$\int_{\chi_{\kappa}} dV = \int_{\kappa} J \, dV.$$

Recall now that coordinates of a point X in the reference configuration are $\mathbf{X}$ and in the current configuration are $\mathbf{x}(t)$. The function J, thus, is the Jacobian (*i.e.*, the determinant of the Jacobian matrix[8]) of the coordinate transformation. This should be expected, since the Jacobian just gives the relative change of the infinitesimal volume due to the coordinate transformation.

2.4. CONSERVATION LAWS

The main trend in all developed natural sciences is a construction of axiomatic theories, *i.e.*, an attempt of reduction of the totality of accumulated knowledge about the subject to a small set of postulates. The rest, on a plan, should be derived from them by the rules of the formal logic. Essentially, these postulates are the model of the object, since all other information about the object must be their necessary consequence. Otherwise, the set of postulates is incomplete.

Fluid mechanics has achieved a significant progress in this regard. Many different motions of liquids and gases can be described by a system of several differential equations. One may not but marvel at such a concentration of knowledge: the number of equations in the system and the number of its different solutions are incommensurable. These differential equations, in turn, are obtained under some assumptions, as consequences, from the integral laws of motion of the continuum.

2.4.1. Integral Conservation Laws

The choice of integral laws as the basis is clear: just integral relations may be obtained from observational data, since only integral parameters are available for measurements. Some of these relations, having the most general character, receive the status of the *laws of nature*[9]. From the mathematical point of view the law of nature is equivalent to the concept of axiom, base assertion accepted without proof. Usually they are formulated either as

1. *conservation laws* of values of certain integral parameters under the fluid motion, or as

2. *balance relation*s: changes of values of integral parameters in a volume are caused (are balanced) by production inside and inflow from the outside.

Consider the first option and its consequences.

One of the fundamental principles of the modern natural science is the so-called *causality principle* according to which any effect (change) must have a cause, and effect cannot precede its cause. If somehow the causes of change of some integral parameter $\Pi(t, \mathcal{B})$ became known, this may be written down as the equation

$$d_t \Pi = \Sigma. \tag{2.10}$$

The quantity Σ describing the rate of change of Π under the action of the cause, is called the *power of source* (*sink*) of this parameter. In the absence of cause, *i.e.*, when the power of source/sink is zero $\Sigma = 0$ the values of $\Pi(t, \mathcal{B})$ do not change over time

$$d_t \Pi = 0. \tag{2.11}$$

In this case, the parameter Π is the quantity which conserves in time. Both equations (2.10) and (2.11) are called the *integral conservation laws*.

2.4.2. Differential Conservation Laws

Since the integral parameter Π satisfies the equalities (2.10) or (2.11), its density π must also satisfy certain relations, called *differential conservation laws*. To derive these relations we write the power of source Σ as

$$\Sigma(t, \mathcal{B}) = \int_{\chi} \sigma(t, \mathbf{x}(t)) dV. \tag{2.12}$$

The density $\sigma\,(t, \mathbf{x}(t))$ is called *generation* (*production*) of Π.

Now the differential equation corresponding to the conservation law (2.10), may be found *via* substitution of expressions (2.3) and (2.12) into (2.10). This gives

$$d_t \int_\chi \pi dV = \int_\chi \sigma dV. \tag{2.13}$$

The scheme of the further reasoning is as follows. We calculate the time derivative and then try to get equality of the kind

$$\int_\chi A(\pi)dV = 0, \tag{2.14}$$

where $A(\pi)$ is a function of density. Assuming, further, the continuity of the integrand and arbitrariness of configuration, we find the required relation[10] $A(\pi)=0$. However, there is a difficulty here, due to the fact that configuration χ generally depends on time and, therefore, operators of differentiation and integration in (2.13) do not commute, *i.e.*, the result depends on their order. They cannot simply swap places. To overcome this difficulty, we use the, so-called, *transfer theorem*[11], which now we will formulate and prove.

2.4.3. Transfer Theorem

Let $\chi = \chi(t, \mathcal{B})$ be a configuration of the body $\mathcal{B}$ corresponding to time t, and $\pi(t,\mathbf{x}(t))$ be a function defined on points $\mathbf{x} \in \chi$. Each configuration is a cross-section of the world-tube $\lambda(t, \mathcal{B})$ of the body by a space of simultaneous events, *i.e.* space of places, associated with a particular moment of time. The chosen frame of reference ϕ includes coordinates on the t-axis and coordinate system in the space of places.

In the general case, points of the body, in the space of places, move non-uniformly relative to the selected coordinate system. As a result, the world-tube of the body is curved, and configurations of the body at different times are different, *i.e.*, are time-dependent. If not for this dependence, the operator $d_t(\cdot)$ might be inserted under the integral sign and the problem would disappear. The curvature of the world-tube manifests itself only in an *inertial frame of reference*[12]. Selecting a particular non-inertial frame of reference, one may achieve straightness of any world-tube.

Let's try to take advantage of this observation. Consider in a space of places a coordinate transformation S which depends on time such that images of the world-lines in $\mathbb{R}^4$, and hence, images of the world-tube under the action of S

become straight (Fig. **2.3**). Cross-sections of the world-tube by spaces of simultaneous events in new coordinates are equal to each other in the sense that they are always consist of the same points of the space $\mathbb{R}^3$, and points of the body always have the same spatial coordinates, regardless of time. The rectifying transformation S must be a one-to-one mapping, otherwise we would not be able to return back. However, this requirement does not define it uniquely. Indeed, there is a lot of ways to straighten a world-tube. Usually do so. The reference configuration κ is selected according to (2.6), and coordinate systems in spaces of places P_t, which correspond to different moments of time are chosen such that coordinates of points over time do not change (Fig. **2.4**). The configuration of the body, in this case, will not change also, remaining equal to $\kappa(\mathcal{B})$.

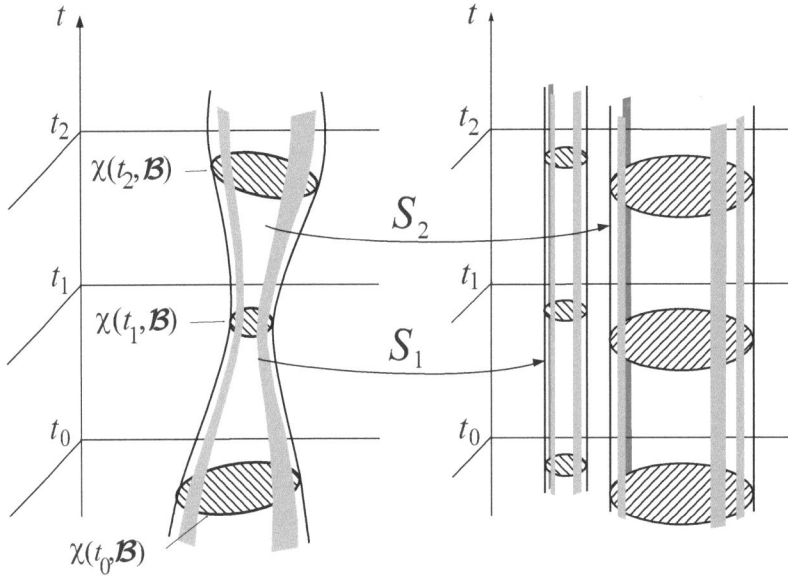

Fig. (2.3). World-tube and options for its rectification.

As a result of application of such transformation S the integral in the left side of (2.13) reduces to the integral over the reference configuration and the integrand is multiplied by the Jacobian J of corresponding transformation of the Lagrangian coordinates to the Euler coordinates $J = |\partial_{x_i} x_j|$. This is the usual change of variables in integrals. Now the left side of (2.13) is rewritten in the form

$$d_t \int_\kappa \pi J \, dV, \qquad\qquad (2.15)$$

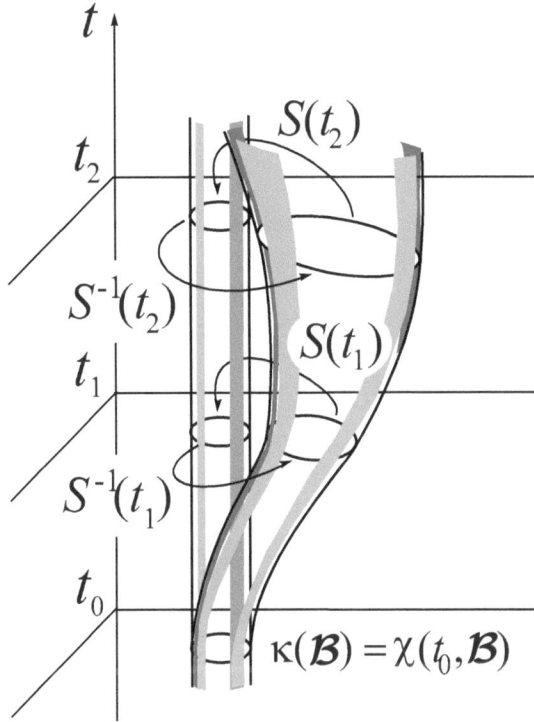

Fig. (2.4). On the proof of the transfer theorem: rectification of the world-tube.

and, since the configuration κ is independent of time, the operator of differentiation can be inserted under the integral sign and we obtain

$$\int_\kappa d_t(\pi J)\,dV. \tag{2.16}$$

In order to write (2.13) in the form (2.14), it is necessary either to express (2.12) in terms of the reference configuration, or (that's what we shall do) to return back to the current configuration χ in (2.16), using the inverse transformation S^{-1}. Such transformation always exists, because transformation S was chosen as a one-to-one mapping of the Euler coordinates in the Lagrangian ones. Integrand is again multiplied by the Jacobian of the coordinate transformation. But now this is the reverse transformation, and hence the Jacobian of this transformation is equal to $\frac{1}{J}$.

One has

$$\int_\chi \frac{1}{J} d_t(\pi J) dV. \tag{2.17}$$

As a result, we obtain the relationship between the rate of change of the integral parameter with the rate of change of its density:

$$d_t \int_\chi \pi dV = \int_\chi \frac{1}{J} d_t(\pi J) dV. \tag{2.18}$$

The *transfer theorem* asserts that the equality (2.18) holds. This is a kinematic theorem, and it holds for any density π, regardless of its nature.

2.4.4. Differential Conservation Laws (Continuation)

Applying the transfer theorem (2.18) to the expression (2.13), we immediately obtain the equation of the type (2.14)

$$\int_\chi \left(\frac{1}{J} d_t(\pi J) - \sigma \right) dV = 0.$$

This implies that the integral conservation law (2.11) holds in case the density π satisfies the equation

$$\frac{1}{J} d_t(\pi J) = \sigma. \tag{2.19}$$

This equation is called the *differential conservation law* corresponding to the integral conservation law (2.11). Differentiating the product πJ we get another form of the same equation

$$d_t \pi = \sigma - \pi \frac{1}{J} d_t J. \tag{2.20}$$

What do these expressions mean? The functions π and J are defined at points $(t, \mathbf{x}(t))$ of the space of events, which together form the world-lines $\lambda(t, X)$ of the points of the body $X \in \mathcal{B}$. Each world line is parameterized by time $t \in \mathbb{R}^1$, and both functions π and J along any world line λ depend only on the values of the parameter, *i.e.* time. Differential conservation law (2.20) asserts that the rate of change of π along any world line is determined by the combined action of generation σ and the relative change of the volume. At the same time, if generation is absent, then according to (2.19), any change in the density of an integral parameter is inversely proportional to the change of infinitesimal volume.

The function π is the density of the integral parameter II and J is equal to relative expansion ($J > 1$) or compression ($J < 1$) of the infinitesimal volume moving with the body. If an isolated body during its motion is compressed and the volume of its configuration is reduced, then $J < 1$ and the density π increases just enough so that the product πJ at a given point of the body (*i.e.*, on the same world-line) has retained its previous value.

Let us integrate the equation (2.19), assuming $\sigma = 0$. The general solution is $J\pi = $ const, *i.e.*, infinitely many particular solutions exist. In order to select the only one solution we are interested in, we should specify some additional condition, which this solution should satisfy. For example, it is possible to specify the value of the function (solution) at some point. As a rule, the initial time moment t_0 is chosen. This gives the condition its name — the *initial condition*. Since J is the Jacobian of coordinates transformation (Lagrangian to Euler), in the initial moment of time when the current configuration coincides with the reference one, $J = 1$. Using the notation $\Pi|_{t=t_0} = \pi_0$ we obtain the following problem:

$$d_t(\pi J) = 0, \tag{2.21}$$

$$J\pi|_{t=t_0} = \pi_0, \tag{2.22}$$

which is called the *initial value problem* or the *Cauchy problem*[13]. We are interested in a particular solution $J\pi = \pi_0$. It coincides with the initial condition, but is valid for any moment of time. Therefore, the value of density π_0 connected with a point of the body in the reference configuration conserves along the world-lines.

Thus, each conserving[14] characteristic of a body is associated with integral conservation law, and corresponding differential conservation law. Further we shall consider other formulations of the differential conservation law (2.19) and will link it with the *balance equation*, which is close to it in the meaning.

NOTES

[1] Taylor Brooke (1685–1731), an English mathematician.

[2] For example, the length of the free path of molecules in a substance or the average distance between stars in a star cluster may be considered as such distance.

[3] Knudsen Martin Hans Christian (1871–1949), a Danish physicist and oceanographer.

[4] This explains the widespread use of the concepts and methods of the theory of continuum in such areas of physics, where there is nothing similar to continuous medium, say, in astrophysics (see, *e.g.*, [3]). Despite huge distances between celestial bodies (from the human's viewpoint), the spatial scales of the studied objects are so grandiose, that the inequality Kn < 0.1 holds.

[5] The set is called *connected* if every pair of its points may be connected by a continuous curve lying in this set, *i.e.*, consisting of points of the same set.

[6] Named after a French mathematician and mechanic Joseph Louis Lagrange (1736–1813).

[7] Euler Leonard (1707–1783), mathematician, physicist and astronomer. He was born and educated in Switzerland, worked in Germany and Russia.

[8] Jacobi Carl Gustav Jacob (1804-51), a German mathematician.

[9] Pay attention to the epigraph to this part of the book.

[10] An integral over an arbitrary domain of a continuous function is equal to zero if and only if this function is identically zero.

[11] This theorem occurs under different names. The name used here is borrowed from the book [4].

[12] *Galilean principle of relativity* postulates the existence of frames of reference

which are called *inertial* and have the following properties: 1) in all inertial frames of reference the laws of nature are the same and 2) frame of reference moving uniformly with respect to an inertial frame of reference is inertial.

[13] Cauchy Augustin Louis (1789–1857), a French mathematician.

[14] In case of absence of external influences.

CHAPTER 3

Rates of Change of Characteristics of Continuum

Abstract: Methods of description of the rates of change of characteristics of a medium are studied. The rate of change of a place of a point of a body is considered. The rates of change of characteristics specified by scalar and vector functions are also discussed. The index summation convention is introduced.

Keywords: Christoffel symbol, Covariant derivative, Dummy index, Euler's formula, Free index, Index summation convention, Parallel transfer, Rate of change of location, Rate of change of scalar function, Rate of change of vector function, Total derivative, Trajectory.

3.1. THE RATE OF CHANGE OF LOCATION: TRAJECTORIES OF MOTION

Simplicity with which the equation (2.19) was integrated, is largely illusory, and the formula (2.21) has a theoretical, but not a practical value. The thing is that the equation (2.19) describes the change of the quantity (πJ) along the world line. Accordingly, the formula (2.21) holds along the same set of points. But this very set is unknown! How to find it?

Recall that a set of places, each of which by turns is occupied at different times by the point X of the body, may be represented as a curve λ permeating a "pile" of spaces of places P_t. This curve is the world-line of a point of a body. If we project all spaces P_t onto one of them (anyone) P, the world-line will be mapped on the three-dimensional curve $\chi(t, X)$, called the trajectory, and vectors $\vec{u}(t)$ tangent to the world-line will be mapped on vectors $\vec{v}(t)$ tangent to the trajectory χ. Since these two totalities of vectors are connected by relation $\vec{u} = (1, \vec{v})$, knowing one of them, we always may calculate the other.

If a family of trajectories is specified, then at any point of each trajectory a tangent vector $\vec{v}$ by definition is equal to

$$\vec{v} = d_t \chi(t, X). \tag{3.1}$$

This expression defines the velocity of motion along the trajectory χ. And on the contrary, if at each point of some region of the space of places, vector $\vec{v}$ is specified, *i.e.*, a velocity field[1] $\vec{v}$, is defined, then the expression (3.1) may be regarded as equation describing trajectories of the moving points of the body, corresponding to a given field. If a coordinate system is defined in the space of places, the equation (3.1) can be written in the form

$$\vec{v} = d_t \mathbf{x}(t). \tag{3.2}$$

The general solution of (3.2) describes all possible trajectories of motion for a given velocity field. In order to select a single particular solution from this infinite totality (general solution) it is necessary to specify the initial condition, *i.e.*, to pose the Cauchy problem:

$$d_t \mathbf{x} = \vec{v}, \tag{3.3}$$

$$\mathbf{x}\big|_{t=t_0} = \mathbf{X}. \tag{3.4}$$

Here (3.4) is the initial condition.

To perform calculations, it is necessary to write down this problem in the so-called component form. This means the following. If some vectors (in our case they are $d_t\mathbf{x}$ and $\vec{v}$) satisfy a certain relation (here, equation (3.3)), then corresponding relations should also be valid for components of these vectors. Let at some spatial point, we are interested in, a coordinate basis $\{\vec{e}_i\}_{i=1}^{3}$ is specified. With respect to this basis both vectors in the equality (3.3) are defined by their components and may be written as follows:

$$d_t \mathbf{x} = (d_t x_1)\vec{e}_1 + (d_t x_2)\vec{e}_2 + (d_t x_3)\vec{e}_3, \qquad \vec{v} = v_1\vec{e}_1 + v_2\vec{e}_2 + v_3\vec{e}_3.$$

Writing down both vectors, say in the left-hand side of (3.3), we factor out common basis vectors and obtain

$$d_t \mathbf{x} - \vec{v} = (d_t x_1 - v_1)\vec{e}_1 + (d_t x_2 - v_2)\vec{e}_2 + (d_t x_3 - v_3)\vec{e}_3 = 0. \tag{3.5}$$

All these operations are possible only if $d_t x$ and $\vec{v}$ are elements of a vector space (recall the axioms of a vector space written out on p.15). We have just such a case: two identical vectors of the tangent vector space defined at the point of the space of places P_t with coordinates $\mathbf{x}$. The right side of (3.5) is the zero vector, all components of which, by definition, are equal to zero. Hence, the left-hand side of the equality is the zero vector too and its components are also equal to zero. Therefore, for $i = 1, 2, 3$ we obtain the equalities

$$d_t x_i = v_i,$$

which hold simultaneously. In other words, one vector equation corresponds to a system of equations for its components. By similar reasoning, we derive the required initial conditions

$$x_i|_{t=t_0} = X_i.$$

However, instead of a tangent vector space the vector space of the radius-vectors should be considered here. The basis (unit) radius-vectors $\{\mathbf{e}_i\}_{i=1}^3$ together with all other radius-vectors begin at some common point (most often the origin of the Cartesian coordinate system is used). Note that the space of radius-vectors, one of which is the point $\mathbf{x}$, and a space of tangent vectors at the same point $\mathbf{x}$ are different vector spaces (see Fig. **3.1**).

Positions of points in an arbitrary (current) configuration are described by their Euler coordinates, changing in time relative to a given coordinate system. In the initial time moment, the configuration of the body by definition coincides with the reference one, and coordinates of points of the body coincide with their Lagrangian coordinates. Thus, each partial solution of the problem (3.3 , 3.4) describes the trajectory of motion of the point X of the body marked with Lagrangian coordinates $\mathbf{X}$, and gives, thereby, the connection between the Euler and Lagrangian coordinates for the same point of the body.

3.2. THE RATE OF CHANGE OF A SCALAR FUNCTION: INDEX SUMMATION CONVENTION

Not only coordinates of a place of a point of a body may change during the motion, but other characteristics as well. For example, densities of integral parameters, defined by scalar fields, or the rates of flow of various processes (change of places including) defined by vector fields, *etc.* At first, consider the case of a scalar function.

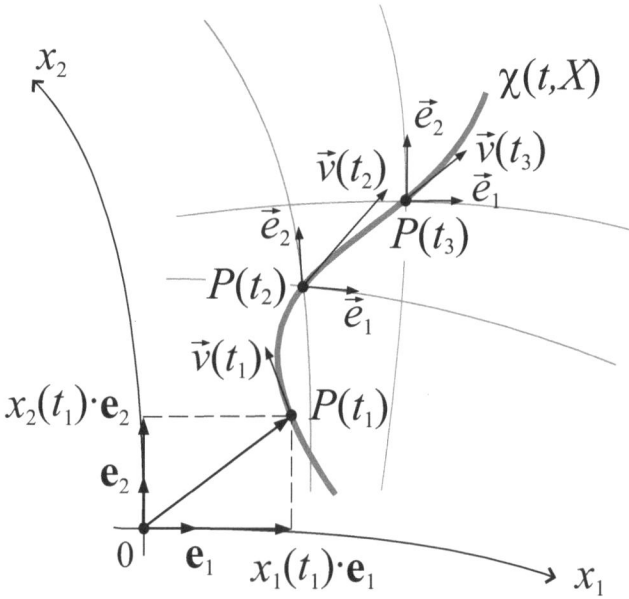

Fig. (3.1). An example of different bases used in description of radius-vectors and tangent vectors. The radius-vector of a point $P(t_1) = x_1(t_1)\mathbf{e}_1 + x_2(t_1)\mathbf{e}_2$ on the trajectory $\chi(t, X)$ is described using the basis $\{\mathbf{e}_1, \mathbf{e}_2\}$. Coordinate lines of the given coordinate system, connected with arbitrary point, say, $P(t_2)$ or $P(t_3)$ of the trajectory, give birth to a coordinate basis at this point. Each tangent (velocity) vector, *e.g.*, $\vec{v}(t_1)$, $\vec{v}(t_2)$ or $\vec{v}(t_3)$, is described here using corresponding coordinate basis $\{\vec{e}_1(t_i), \vec{e}_2(t_i)\}$.

Let some scalar field a be defined at points of a body, *i.e.* at any time t each point X of the body, with coordinates $(t, \mathbf{x}(t))$ in the space of events, is associated with a value of the scalar function $a(t, \mathbf{x}(t))$. The point X is moving over time along its world-line. At that, the values of the function a are changing with the rate $d_t a$. In the general case this rate of change is equal to the time derivative of a composite function of time, namely:

$$
\begin{aligned}
d_t a(t, \mathbf{x}(t)) &= d_t a(t, x_1(t), x_2(t), x_3(t)) \\
&= \partial_t a + \sum_{i=1}^{3} (\partial_{x_i} a) d_t x_i \\
&= \partial_t a + \sum_i v_i \, \partial_{x_i} a \\
&= \partial_t a + v_i \, \partial_{x_i} a.
\end{aligned}
\tag{3.6}
$$

For brevity the last expression is written using the so-called *index summation convention*. It lies in the fact that when writing the sum of a series, in case the general term contains a pair of indices of summation it is possible to omit the symbol of summation, and use this pair of repeated indices as an indicator of required summation. Such indices are called *dummy indices*. They do not appear in the result of summation and, therefore, may be labeled with any icon not occurring in this expression.

The case when summation is not supposed, has to be mentioned specially. So, if the general term of a series is discussed, it is clear that there is no question about summation. Furthermore, not every series may be written using this rule. Usually it is used in those expressions, which regularly meet with series with the same number of like terms. The situation is typical in continuum mechanics. Expressions, formidable due to their complexity and length, miraculously transform and become easily visible with the application of the summation rule. We are going to use it widely.

Whenever found the expression like $a_i b_{jk} c_{kjin}$, it means that this is a triple series with general term of the above form, *i.e.*, $\sum_i \sum_j \sum_k a_i b_{jk} c_{kjin}$. Note that the index n does not have a pair, and hence no summation over this index is performed. Such indices are called *free indices*. Expression, which contains free indices, is valid for all values of their prespecified set (in our case these values are 1, 2, 3).

In a more complex case, when the general term of a series is not a monomial, a series must be written as the sum of series with monomial general terms. For example, the expression $\sum_i a_i(b_{ij} + b_{ji})$ we first rewrite as $\sum_i a_i b_{ij} + \sum_i a_i b_{ji}$, and then, applying the summation rule, we obtain $a_i b_{ij} + a_i b_{ji}$. Here, each term has the same free index j which marks all terms as they were labeled in the original expression. Of course, the index itself may be denoted arbitrarily, but its designation should be the same in all summands. On the contrary, dummy indices in each term may be designated differently, because the equality

$$\sum_i a_i(b_{ij} + b_{ji}) = \sum_i a_i b_{ij} + \sum_k a_k b_{jk} = a_i b_{ij} + a_k b_{jk},$$

in any case, holds. The need for such redesignation of dummy indices often occurs when converting and simplifying expressions. If, for instance, the expression $a_i b_{ij} + a_k b_{jk}$ is obtained, then it may be convenient to denote the dummy index in the last term by the letter i (or, conversely, use the letter k in the first term) in order to factor out the common factor   and   write   either $a_i(b_{ij} + b_{ji})$ or $a_k(b_{kj} + b_{jk})$.

Let us return to the formula (3.6). We are going to interpret the numbers $\partial_{x_i} a$ as components of a vector, which will be denoted as $\nabla a = (\partial_{x_1} a, \partial_{x_2} a, \partial_{x_3} a)$ and called the *gradient* of a scalar field a at some point $\mathbf{x}(t)$. At any given time, this vector characterizes the spatial inhomogeneity of the field a and is directed towards its maximum growth. Quite often it is possible to meet designation of a gradient by the symbol grad. Thus,

$$\nabla a = \operatorname{grad} a = (\nabla a)_i \vec{e}_i = \left(\partial_{x_i} a\right)\vec{e}_i. \tag{3.7}$$

If the last term in (3.6) is regarded as a scalar product[2] $(\vec{v}, \ \nabla a)$, then the rate of change of a function a may be written in the form

$$d_t a = \partial_t a + (\vec{v}, \nabla a). \tag{3.8}$$

What does the formula (3.6) or (3.8) mean? Let us analyze a simple case of two-dimensional space of places (it is easier to draw it). In this case the formula (3.6) will take the form

$$d_t a = \partial_t a\big|_{x_1, x_2} + v_1 \partial_{x_1} a\big|_{t, x_2} + v_2 \partial_{x_2} a\big|_{t, x_1}.$$

Note that partial derivatives at the right side, according to their definition, are calculated for fixed values of all arguments not involved in the differentiation (in the formula, they are explicitly indicated at each derivative). Substituting ratios of finite increments for derivatives, one obtains the approximate equality

$$\frac{\Delta a}{\Delta t} = \frac{\Delta a}{\Delta t}\bigg|_{x_1, x_2} + \frac{\Delta x_1}{\Delta t}\frac{\Delta a}{\Delta x_1}\bigg|_{t, x_2} + \frac{\Delta x_2}{\Delta t}\frac{\Delta a}{\Delta x_2}\bigg|_{t, x_1}.$$

Further, multiplying by Δt and reducing, one may write

$$\Delta a = \Delta a\big|_{x_1, x_2} + \Delta a\big|_{t, x_2} + \Delta a\big|_{t, x_1}. \tag{3.9}$$

The left side of (3.9) is equal to the change of value of the function $a(t, \mathbf{x})$, which corresponds to the shift of a point X along its world-line on Δt from point B to point E' (Fig. **3.2**). The right side is the sum of increments of a, which corresponds to the shift between the same points, but using another route, namely, shifts along lines parallel to the coordinate axes. Various routes correspond to different orders of terms in the right side of (3.9). The formula states that the increment Δa does not depend on the route. The scalar product $(\vec{v}, \nabla a)$ is a projection of ∇a on the direction of movement, *i.e.* $\vec{v}$, and the vector ∇a is directed towards the maximal changes of the function a.

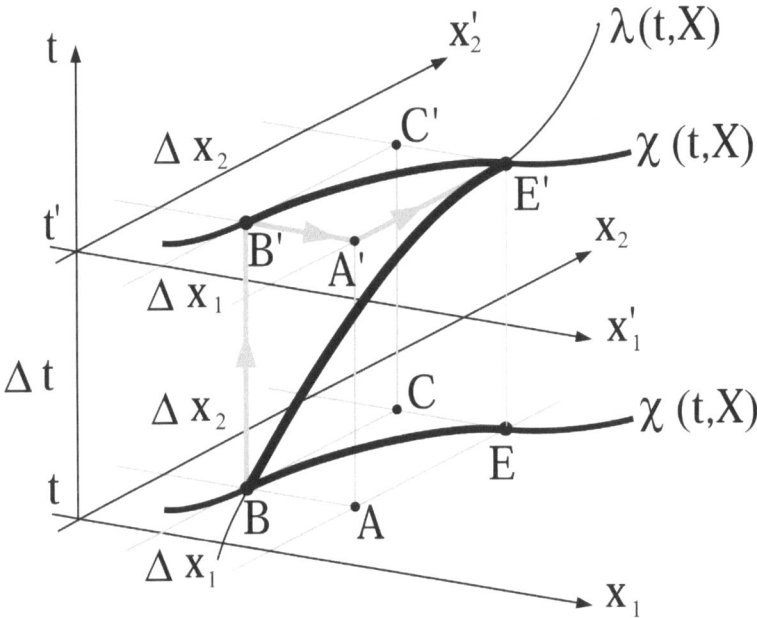

Fig. (3.2). Relation between total $d_t(\cdot)$ and partial $\partial_t(\cdot)$ time derivatives. The route indicated by arrows corresponds to the order of terms in the formula (3.9). Both curves $\chi(t, X)$ are the same trajectory of the point X *i.e.* the projection of the world line $\lambda(t, X)$ onto the space of places. In one case, this space is P_t and $P_{t'}$ in the other one.

Once again we point out the difference between time derivatives[3] on the right $(\partial_t a)$ and left $(d_t a)$ sides of the equality (3.8). The partial derivative with respect to time, like any other partial derivative is calculated at fixed values of all arguments except one (here, time). Derivative $d_t a$, which is called the *total derivative*, makes sense of the rate of change of the function a along the world-line of a point of the body. Synonymous with total derivative is a substantial (or material) derivative, *i.e.*, the derivative associated with the given point of substance (matter).

The formula (3.8) is valid for any scalar function and, therefore, often regarded as a relationship between total time derivative and partial one:

$$d_t(\cdot) = \partial_t(\cdot) + (\vec{v}, \nabla(\cdot)) = \partial_t(\cdot) + (\vec{v}, \nabla)(\cdot), \qquad \textbf{(3.10)}$$

irrespective of any particular function. This relationship is known as the *Euler's formula*.

Exercise. How a world-line should look like, for the equality $d_t(\cdot) = \partial_t(\cdot)$ holds? What kind of motion corresponds to this world line?

3.3. THE RATE OF CHANGE OF A VECTOR FUNCTION

Suppose now that some vector field $\vec{a}$ is defined on points of a body, *i.e.*, each point X of a body with current coordinates $(t, \mathbf{x}(t))$ in the space of events is associated with a vector $\vec{a}(t, \mathbf{x}(t))$. If we imagine a vector as an arrow, this arrow begins at point $(t, \mathbf{x}(t))$. At each point different vectors may be defined. All of them are representatives of those vector spaces, the zero elements of which is this very point. Thus, each point of a configuration of the body is associated with vector spaces, which are "attached" to it by their zero elements.

Now, if we define a tangent vector field on this set of points, we choose vectors one by one from all tangent vector spaces, which are "attached" to the points of this configuration.

As an example we shall discuss the movement of a ball (Fig. **3.3**). Each point of the surface of the ball is moving along its trajectory. Velocity of the motion may be different, but it is always described by a vector, tangent to the trajectory of the point under consideration. Every conceivable velocity vectors of a point of the surface of the ball, form at this point, a tangent vector space. Each point of the surface has its own tangent vector space, and velocity of the point at any time is described by one of the vectors of this space.

Direction of the velocity vector and its length depend on the nature of the movement of the ball. So, if the ball is rotating (Fig. **3.3a**) around a fixed axis passing through its center, the trajectories of all points of the surface are circles lying on the surface of the ball, and the velocity field is a field of vectors tangent to these circles. The direction of the velocity vector depends on position of a point on the trajectory and its magnitude depends on the distance from the axis of rotation. On the contrary, if the ball is moving forward without rotation and acceleration (Fig. **3.3b**), the trajectories of points of the surface are parallel lines.

The velocity field is uniform, *i.e.*, it does not depend on time and position of the point. And finally, if the ball is rolling without sliding along a plane (Fig. **3.3c**), the trajectories of points are trochoids. Each velocity vector is tangent to corresponding trochoid and its length depends on the position of a point on the trajectory.

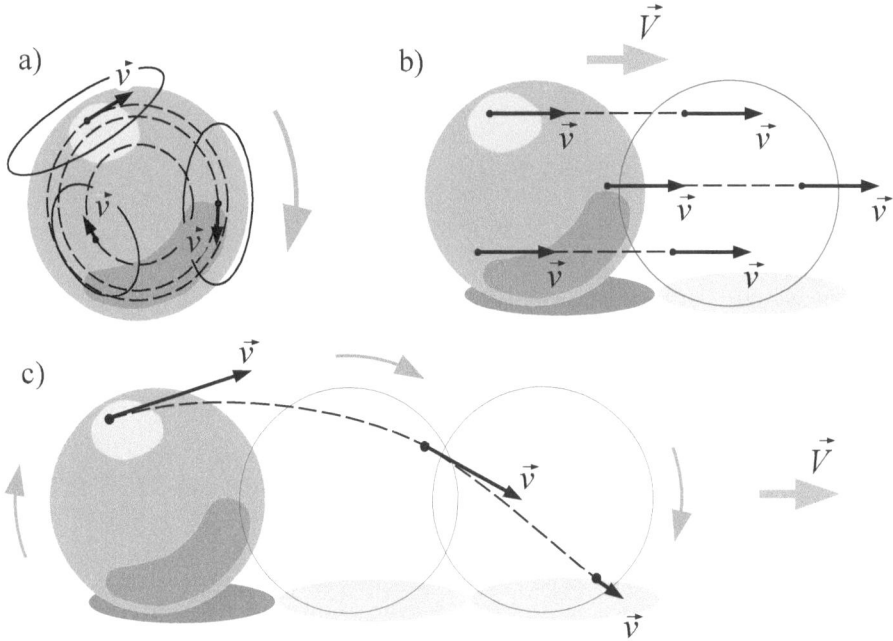

Fig. (3.3). Tangent vector fields on the surface of a sphere: **a**) a ball, rotating around a fixed axis passing through its center; **b**) a ball moving forward without rotation; **c**) a ball rolling on a plane without sliding. Dashed curves are trajectories of the points of the surface. The vector $\vec{V}$ is a translational velocity of the ball, and vector $\vec{v}$, tangent to the trajectory, is the velocity of the point on the surface.

In every vector space it is possible to choose a time-independent basis (let it be the coordinate basis $\{\vec{e}_i(\mathbf{x})\}_{i=1}^{3}$; see, definition on p.10) and provide vectors with corresponding components. Arbitrary vector $\vec{a}$ may be then written as follows

$$\vec{a}(t, \mathbf{x}(t)) = a_i(t, \mathbf{x}(t))\vec{e}_i(\mathbf{x}). \tag{3.11}$$

We will consider *smooth vector fields* only, *i.e.*, those fields, in which corresponding vector components vary smoothly from point to point. This means that if we consider components of vectors of a field as functions of a point (*i.e.*, $a_i = a_i(t, \mathbf{x}(t)) = a_i(t, x_1, x_2, x_3)$), they are differentiable with respect to all their arguments required number of times.

Let us calculate the time derivative of the vector $\vec{a}$ using decomposition (3.11)

$$d_t \vec{a} = d_t(a_i \vec{e}_i) = (d_t a_i)\vec{e}_i + a_i d_t(\vec{e}_i). \tag{3.12}$$

The first term is the vector with components $d_t a_i$ which may be calculated according to (3.11):

$$d_t a_i = \partial_t a_i + v_k \partial_{x_k} a_i = \partial_t a_i + (\vec{v}, \nabla a_i). \tag{3.13}$$

Please note that the dummy index *i* in the formula (3.11) had to be replaced with *k* not to be confused with the number of component of the vector $\vec{a}$ (it is clear, that with the same success a letter-number of component might be replaced, but necessarily in all terms of the expression).

The second term in (3.12) should also be a vector (since, the first one is a vector). It is a linear combination of vectors $d_t \vec{e}_i$. We write each such vector in accordance with the formula (3.6). Given that $\partial_t \vec{e}_i = 0$, one finds

$$d_t \vec{e}_i = v_k \partial_{x_k} \vec{e}_i.$$

On the right side is a linear combination of the vectors $\partial_{x_k} \vec{e}_i$ each of which may be expanded using the basis $\{\vec{e}_j\}$ and written as $\partial_{x_k} \vec{e}_i = \Gamma_{ki}^j \vec{e}_j$. The symbol Γ_{ki}^j here is the standard notation of the expansion coefficients. As a result one obtains

$$d_t \vec{e}_i = \left(\Gamma_{ki}^j v_k\right)\vec{e}_j. \tag{3.14}$$

The quantity $(\Gamma_{ki}^j v_k)$ is the *j*-th component of the vector $d_t \vec{e}_i$ with respect to the above-mentioned basis. In accordance with the summation rule, the right side of (3.14) contains the double sum over *j* and *k*. Finally, collecting together (3.12), (3.13) and (3.14), one finds

$$d_t \vec{a} = \left(\partial_t a_i + v_k \partial_{x_k} a_i\right)\vec{e}_i + \Gamma_{ki}^j v_k a_i \vec{e}_j = \left(\partial_t a_i + v_k\left(\partial_{x_k} a_i + \Gamma_{kj}^i a_j\right)\right)\vec{e}_i. \tag{3.15}$$

Designations of dummy indices i and j in the last term had to change in order to factor out the basis vector. Thus, the i-th component of the vector $d_t\vec{a}$ is as follows

$$(d_t\vec{a})_i = \partial_t a_i + v_k\left(\partial_{x_k} a_i + \Gamma^i_{kj} a_j\right) = \partial_t a_i + v_k \nabla_k a_i, \tag{3.16}$$

where the symbol $\nabla_k a_i \equiv \partial_{x_k} a_i + \Gamma^i_{kj} a_j$ denotes the so-called *covariant derivative* in the direction of the vector $\vec{e}_k$. It differs from the partial derivative in the same direction by an additional term. This addend containing the unknown so far quantities Γ^i_{kj}, is generated by the following problem.

Calculation of a derivative of a vector function is related with calculation of a difference of two vectors, because, according to the standard definition the derivative at a point $(t, \mathbf{x}(t))$ of the world-line $\lambda(t, X)$ would have to be equal to the limit:

$$\lim_{\Delta t \to 0} \frac{\vec{a}(t+\Delta t, \mathbf{x}(t+\Delta t)) - \vec{a}(t, \mathbf{x}(t))}{\Delta t}. \tag{3.17}$$

However, this expression is meaningless, since the difference in the numerator is not defined yet. The thing is that the vectors $\vec{a}(t + \Delta t, \mathbf{x}(t + \Delta t))$ and $\vec{a}(t, \mathbf{x}(t))$ belong to different vector spaces. And, in general, all vectors of any vector field belong to different vector spaces, each one to its own, since all vectors of the field, start at different points.

Exercise. Test yourself: do you understand the difference between a vector space (see p. 15) and a vector field (see footnote on p. 39)?

Thus, it is clear that the rate of change of a vector function cannot be defined by direct generalization of the scalar case. First of all, one should figure out how to compare vectors belonging to different vector spaces. How to move a vector from one vector space to another without change, and what is meant by the lack of change? One method is the so-called *parallel transfer*, using which the vector $\vec{a}(t + \Delta t, \mathbf{x}(t + \Delta t))$ may be moved to the point t on the same world-line (Fig. **3.4**).

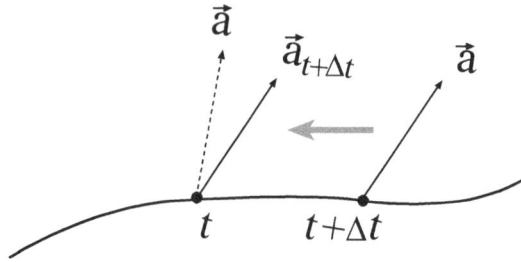

Fig. (3.4). Translation.

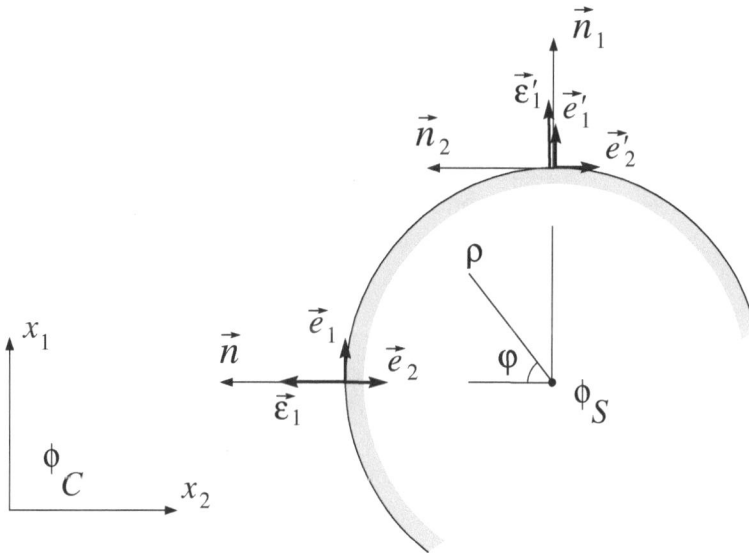

Fig. (3.5). Parallel transport of a normal vector on a sphere.

Unfortunately, the only method of parallel transport of vectors in the general case does not exist, and everything depends on the chosen point of view. To illustrate difficulties that arise here, consider the following example. Let somewhere, say, at the Equator, we have constructed a vector $\vec{n}$ normal to the surface. Now, we move it to the Pole, keeping parallel to itself (see. Fig. **3.5**, where for simplicity two-dimensional image is drawn). What vector at the Pole should be regarded as parallel to the original one: normal $\vec{n}_1$, tangent vector $\vec{n}_2$ or some other?

If at the Equator and at the Pole the coordinate basis $\{\vec{e}_1, \vec{e}_2, \dots\}$ associated with the Cartesian coordinate system ϕ_C is used then the tangent vector $\vec{n}_2$, perhaps, should be regarded as parallel: although it became tangent, its components are the same as the original vector $\vec{n} = (0, n_C, 0)$ has. Furthermore, if we now calculate the time derivative of the vector $\vec{n}$ using the formula (3.17)

$$\lim_{\Delta t \to 0} \frac{\vec{n}_{\parallel}(x(t+\Delta t)) - \vec{n}(x(t))}{\Delta t}, \tag{3.18}$$

where $\vec{n}_{\parallel}(x(t + \Delta t))$ denotes the vector $\vec{n}(x(t + \Delta t))$ transferred parallel to the point $x(t)$, then, obviously, we obtain the zero vector. Such a result may be interpreted as transfer of the vector $\vec{n}$ unchanged, and that is exactly what is required from the parallel transport.

However, another coordinate basis may be used. For example, let the basis $\{\vec{\varepsilon}_1, \ldots\}$ is associated with a spherical coordinate system ϕ_s (the only basis vector, which will be needed, is indicated here). In this case, the vector $\vec{n}_1$ appears to be parallel inasmuch as it is normal, and its components with respect to the new basis are the same as the original vector $\vec{n} = (n_S, 0, 0)$ has. Finally, the time derivative, obtained from the formula (3.18), is also zero. In geophysical problems such point of view is clearly preferable.

Note here that at any point other than the initial, the resulting vectors of these two of "parallel" translations do not coincide, and only one transfer method is considered as parallel transfer from each viewpoint. Accordingly, only one of these two vectors is regarded as parallel, to the initial one.

The selected rules of the parallel transport are described by specifying the coefficients Γ_{ki}^{j} in the formula (3.14). They show how different vector spaces are connected with each other and are called the *coefficients of connexion*, or the *Christoffel*[4] *symbols*, on behalf of the mathematician who described them. This is a kind of additional structure, since vector spaces may exist without indication of such a connection. When a Cartesian basis is used, all Christoffel symbols are zero, and the covariant derivative is equal to the partial. In other coordinate bases, at least some of these symbols are non-zero and should be taken into account when calculating derivatives.

We will not specifically study the covariant differentiation, but in order not to limit ourselves only within the framework of Cartesian coordinates, we will write, wherever needed, the covariant derivatives instead of partial, especially since their properties are very close. For instance, a covariant derivative in the direction of $\vec{e}_k$ of a scalar function is equal to partial derivative $\nabla_k f = \partial_{x_k} f$, and a covariant derivative of a vector $\vec{a}$ is a vector $\nabla_k \vec{a} = (\nabla_k a_i)\vec{e}_i$. In addition, the covariant derivative is linear, *i.e.*, if f is a scalar function and $\vec{a}$ and $\vec{b}$ are vectors, then the equalities

$$\nabla_k(\vec{a} + \vec{b}) = \nabla_k\vec{a} + \nabla_k\vec{b},$$
$$\nabla_k(f\vec{a}) = (\nabla_k f)\vec{a} + f(\nabla_k\vec{a}) = (\partial_{x_k} f)\vec{a} + f(\nabla_k\vec{a})$$

hold. These relations allow one to write (3.6) and (3.16) as the *Euler's formula*, which is valid for any object in arbitrary coordinate system:

$$d_t(\cdot) = \partial_t(\cdot) + v_k\nabla_k(\cdot). \tag{3.19}$$

If we introduce symbolic vector $\nabla = (\nabla_1, \nabla_2, \nabla_3)$, the last term in (3.19) may be written as the scalar product and the whole formula, formally becomes indistinguishable from (3.10). The difference is only in the sense of the symbolic vector (operator) ∇. Further in this course we will assume that coefficients of connexion are given, rules of parallel transport are known, and differential operations are always defined.

NOTES

[1] A *field* of the quantity φ specified in some area V is a rule that assigns each point in V the value of φ at this point [5]. A field is called a scalar field, if φ is a scalar quantity, and it is called a vector field, if φ is a vector, and so on.

[2] The *scalar (or inner) product* of two vectors is a rule $(\cdot, \cdot)$, which associates a pair of vectors $\vec{a}, \vec{b}$ of a given vector space, and a scalar $(\vec{a}, \vec{b})$, *i.e.*, a real number. By definition, it is symmetric $(\vec{a}, \vec{b}) = (\vec{b}, \vec{a})$ and linear in both arguments, *i.e.*, if α and β are constants, and $\vec{c}$ is another vector, then

$$(\alpha\vec{a} + \beta\vec{b}, \vec{c}) = \alpha(\vec{a}, \vec{c}) + \beta(\vec{b}, \vec{c}).$$

If the basis $\{\vec{e}_i\}$ is selected, the scalar product may be written as follows (the summation convention is used here)

$$(\vec{a}, \vec{b}) = (a_i\vec{e}_i, b_j\vec{e}_j) = a_ib_j(\vec{e}_i, \vec{e}_j) = g_{ij}a_ib_j.$$

The totality of numbers $g_{ij} \equiv (\vec{e}_i, \vec{e}_j)$ defines a rule for calculating the scalar product. If we choose these numbers to be equal to

$$g_{ij} = \begin{cases} 0, i \neq j, \\ 1, i = j, \end{cases}$$

then $(\vec{a}, \vec{b}) = a_i b_i$ and the basis with respect to which numbers a_i and b_j are components of the vectors $\vec{a}$ and $\vec{b}$, is called the *orthonormal basis*. We shall use only such bases. Consideration of a more general case may be found in, *e.g.*, [5].

---

[3] As a matter of fact, there is no fundamental difference. In both cases the same definition of derivative is used: the limit of the ratio of increments of a function and an argument. However, since we consider functions of several arguments, to compute the derivative we must somehow get rid of the "extra" arguments. In case of partial derivative $(\partial_t f)$ extra arguments are fixed, the function of several variables turns here into function of one variable t, and the change of this function along one-dimensional set (the coordinate line where only t-coordinate changes) is investigated. In case of total derivative $(d_t f)$ the same effect (transformation of a function of several variables into a function of one variable) is achieved by considering the change of the function along another one-dimensional set: the world-line, which is defined by the velocity field. Since one-dimensional sets, along which the function change is investigated, are generally different, corresponding derivatives are different also $\partial_t f \neq d_t f$.

[4] Christoffel Elvin Bruno (1829-1900), a German mathematician.

CHAPTER 4

Minimum Information About Tensors

Abstract: Minimal necessary information about tensors is presented. Definition of a tensor of the 2nd rank is given, and components of a tensor with respect to some basis are considered. Algebraic and differential operations, allowing to build tensor expressions are discussed. Some special tensors are defined. Also the eigenvalue problem is studied.

Keywords: Antisymmetric tensor, Contraction of tensors, Divergence, Eigenvalue problem, Gradient, Inner product, Inverse tensor, Orthogonal tensor, Symmetric tensor, Tensor, Tensor expression, Tensor operation, Tensor rank, Transposed tensor.

4.1. TENSORS AND TENSOR OPERATIONS

4.1.1. Introduction

In the previous chapter, we first met a new geometric object, a *tensor*. Up to now, we had geometric objects of two types: scalars and vectors. An example of a scalar is the mass density ρ, an example of a vector is the velocity of motion $\vec{v}$.

The mass density is a number. A component of a vector $\vec{v}$ with respect to some basis is a number too. Can this component be considered as scalar? No, it can't. Why then one of these numbers is scalar and the other is not?

The point, of course, is not in numbers, per se, but in the sense that we invest in these or other objects, accepting numerical values. So, ρ makes sense of the mass density and v_i makes sense of component of a vector with respect to the chosen basis. The last indication is decisive: whereas ρ is the mass density regardless to any basis, the quantity v_i loses its meaning of a vector component without reference to the chosen basis. At that, the vector $\vec{v}$ itself, makes sense of the velocity of motion irrespective of the choice of the basis. Quantities ρ and $\vec{v}$ exist by themselves (within this or that problem). Systems of coordinates, bases and the like, are needed only to be able to describe vectors and more complex objects *via* numbers, and to replace the work with these objects by the work with numbers (their components). However, these numbers characterize our objects only with respect to the pre-selected reference (basis) objects, and not by themselves.

By studying the rate of change of a scalar function (field) a, we found that at each point derivatives $\partial_{x_i} a$ are components of the vector ∇a with respect to coordinate basis $\{\vec{e}_i\}$. In the study of the rate of change of the vector function (field) $\vec{a}$ we have obtained the set of quantities $\nabla_k a_i$. With respect to the same coordinate basis, they are components of a new geometric object, a tensor $\nabla \vec{a}$. By analogy with the name of the vector ∇a this tensor $\nabla \vec{a}$ is called the *gradient of a vector field* $\vec{a}$. Due to the fact that $\nabla_k a_i$ is a two-index quantity, it is called a component of a tensor of the 2nd rank. It is convenient to represent a tensor as a matrix of its components

$$\nabla \vec{a} = \begin{pmatrix} \nabla_1 a_1 & \nabla_2 a_1 & \nabla_3 a_1 \\ \nabla_1 a_2 & \nabla_2 a_2 & \nabla_3 a_2 \\ \nabla_1 a_3 & \nabla_2 a_3 & \nabla_3 a_3 \end{pmatrix}.$$

Like other geometric objects, the tensor $\nabla \vec{a}$ does not depend on the choice of the basis, *i.e.* it exists by itself.

Not all quantities provided with indices are components of tensors. For instance, the Christoffel symbols do not form a tensor. For details see, *e.g.*, [5].

4.1.2. Tensor of Rank 2 and its Components

So, what is a tensor? There are different ways of thinking about this concept[1]. We will use the following definition.

A tensor of rank 2 is a linear mapping of a vector space onto itself.

Thus, a tensor is a linear transformation of vectors of some vector space to vectors of the same vector space. In other words, a tensor is a linear vector-valued function defined on vectors.

For example, if $\vec{a}$ and $\vec{b}$ are vectors of a vector space and T is a tensor mapping $\vec{a}$ to $\vec{b}$ then

$$\vec{b} = \mathrm{T}\vec{a}. \tag{4.1}$$

Linearity, as usually, means that for each mapping T, every vector $\vec{a}$ and $\vec{b}$ and arbitrary numbers α and β the equality:

$$T(\alpha\vec{a} + \beta\vec{b}) = \alpha T\vec{a} + \beta T\vec{b}$$

holds, *i.e.* it is possible to remove brackets and to factor numerical multipliers out.

Having defined a basis $\{\vec{e}_i\}$ in a vector space, we associate each vector with a set of numbers (its components with respect to the basis): $\vec{a} = a_i\vec{e}_i$ and $\vec{b} = b_j\vec{e}_j$. At that, tensors, which transform elements of the vector space, are provided with corresponding totalities of components. Let's substitute the decomposition of the vector $\vec{a}$ with respect to the basis in (4.1) and obtain

$$T\vec{a} = Ta_i\vec{e}_i.$$

Due to linearity of the tensor it is possible to factor out vector components

$$T\vec{a} = a_i \overbrace{(T\vec{e}_i)}^{\vec{T}_i}.$$

From the definition of a tensor, the expression in parentheses, is some new vector $\vec{T}_i$. Like any other vector it may be written as a decomposition with respect to the basis $\vec{T}_i = T_{ji}\vec{e}_j$. Then the chain of equalities takes the form

$$T\vec{a} = \overbrace{a_i T_{ji}}^{b_j}\vec{e}_j = b_j\vec{e}_j.$$

In the last equality we have introduced a new notation

$$a_i T_{ji} \equiv b_j.$$

Components (here a_i and b_j) unambiguously define vectors (here $\vec{a}$ and $\vec{b}$) with respect to the basis. Therefore, the quantities T_{ji} also unambiguously define the transformation (4.1), *i.e.* the tensor T. They are called the *components of the tensor* T *of rank 2 with respect to the basis* $\{\vec{e}_i\}$. The number of these quantities is equal to the square of dimension of the vector space, on the elements of which the tensor is defined. In a component form it is convenient to write tensor as a square matrix. Now, if the vectors $\vec{a}$ and $\vec{b}$ are written as columns, the expression (4.1) takes the form

$$\begin{pmatrix} b_1 \\ b_2 \\ b_3 \end{pmatrix} = \begin{pmatrix} T_{11} & T_{12} & T_{13} \\ T_{21} & T_{22} & T_{23} \\ T_{31} & T_{32} & T_{33} \end{pmatrix} \begin{pmatrix} a_1 \\ a_2 \\ a_3 \end{pmatrix}$$

and for j-th component of the vector $\vec{b}$ one obtains $b_j = T_{ji} a_i$. This product is called the *contraction* of a vector and a tensor and in component-free form is written as $T\vec{a}$.

All considered geometric objects such as scalars, vectors, tensors of rank 2, may be treated as tensors of various ranks. The rank of a tensor corresponds to a number of indices of its components. So, a scalar is a tensor of the zeroth rank, and a vector is a tensor of the first rank. Since tensors are quantities, which exist independently of the choice of basis, they are convenient means of writing down physical laws. Equations of physical models, written in a tensor form, are compact and are valid in any coordinate system and for any choice of the basis.

4.1.3. Algebraic Tensor Operations

Mathematical operations with tensors, which produce a tensor, are called *tensor operations*. Here is a list of algebraic tensor operations.

1. *Summation* of tensors of the same rank; is performed component-wise and the result is a tensor of the same rank:

$$(\vec{a} + \vec{b})_j = a_j + b_j,$$
$$(A + B)_{ij} = A_{ij} + B_{ij}.$$

2. *Multiplication* of a tensor by number; is performed component-wise and the result is a tensor of the same rank:

$$(\alpha \vec{a})_j = \alpha a_j,$$
$$(\alpha A)_{ij} = \alpha A_{ij}.$$

3. *Contraction* of tensors by a pair of indices; the rank of operands are not less than 1, and the rank of the result is two units less than the sum of ranks of contracted tensors:

$$(\vec{a}, \vec{b}) = a_j b_j,$$
$$(A\vec{a})_i = A_{ij} a_j,$$
$$(AB)_{ij} = A_{ik} B_{kj}.$$

Thus, the result of contraction of two vectors is a scalar (the contraction of two vectors is their scalar product), the contraction of a tensor of rank 2 with vector is a vector and contraction of two tensors of rank 2 is a tensor of the 2nd rank.

4. *Inner product* is defined for tensors of the same rank; the result is a scalar. The inner product of two vectors is often called their scalar product. We shall denote it by

$$(\vec{a}, \vec{b}) = a_j b_j,$$

and the inner product of two tensors of the second rank we shall denote by

$$A : B = A_{kj} B_{jk}.$$

Obviously $A : B = \text{tr}(AB)$, where $\text{tr}(\cdot)$ is the trace of corresponding matrix.

5. *Tensor product*; the rank of the result is equal to the sum of ranks of multipliers:

$$(\vec{a} \otimes \vec{b})_{jk} = a_j b_k.$$

If we limit ourselves to tensors of the 2nd rank then it is easy to see that all the above-listed tensor operations are well-known actions with vectors and square matrices: summation, multiplication by a number, the product of matrices. The inner product of two tensors of the 2nd rank is equal to the trace of the product of two matrices. A familiar example of the fifth operation is the product of a vector-column and a vector-line. The result is a square matrix.

Note that all these algebraic operations are performed at a point where corresponding tensors are defined.

4.1.4. Differential Tensor Operations

Beside algebraic tensor operations there may be others. In case a tensor field is given, it is possible to define the so-called *differential tensor operations*. Here they are.

1. *Gradient*. The rank of the result is by one unit greater than the rank of an initial tensor. For example,

- gradient of a scalar function φ is a vector of the form

$$\nabla\varphi = (\partial_{x_i}\varphi)\vec{e}_i.$$

- gradient of a vector-valued function $\vec{a} = a_i\vec{e}_i$ is a tensor $\nabla\vec{a}$ of rank 2 with following components:

$$(\nabla\vec{a})_{kj} = \nabla_j a_k$$

with respect to the basis $\{\vec{e}_i\}_{i=1}^3$. If the basis is Cartesian, covariant derivatives are equal to partial and the previous expression is written as:

$$(\nabla\vec{a})_{kj} = \partial_{x_j} a_k.$$

2. *Divergence.* The rank of the result is by one unit less than the rank of an initial tensor. For example,

- divergence of a vector $\vec{a} = a_i\vec{e}_i$ is the following scalar

$$div\vec{a} = \nabla_i a_i,$$

- divergence of the 2nd rank tensor T is a vector with components

$$(div\text{T})_j = \nabla_k \text{T}_{jk}.$$

If the basis is Cartesian, covariant derivatives in both cases are equal to partial derivatives. Divergence of a tensor, is often considered as a contraction with the symbolic vector ∇. In case of a vector (a tensor of rank 1), such contraction is equivalent to the scalar product, which is often used when writing the divergence of a vector, namely

$$div\vec{a} = (\nabla, \vec{a}).$$

Here, however, we must keep in mind that ∇ is the differential operator, not a real vector. For this reason, the expression $(\nabla, \vec{a})$ loses one of the important properties of the scalar product, its symmetry, *i.e.*,

$$(\nabla, \vec{a}) \neq (\vec{a}, \nabla).$$

Indeed, the definition of divergence implies that $(\nabla, \vec{a}) = \nabla_i\, a_i$ is a scalar (divergence of the vector $\vec{a}$), while the expression $(\vec{a}, \nabla) = a_i\, \nabla_i$ is the differential operator. Accordingly, the expression $(\nabla, \vec{a})\vec{b}$ is the product of the vector $\vec{b}$ and the number $(\nabla, \vec{a})$, whereas the expression $(\vec{a}, \nabla)\vec{b}$ is equal to the contraction of the 2nd rank tensor $\nabla\vec{b}$ (this is the gradient of the vector $\vec{b}$) with the vector $\vec{a}$ and also may be written in the form $(\nabla\vec{b})\vec{a}$.

Another matter if an expression like $(\vec{u}, \nabla c)$ is used. Here both factors in the scalar product are valid vectors ($\vec{u}$ and the gradient ∇c of some scalar function c) and the symmetry holds, *i.e.* $(\vec{u}, \nabla c) = (\nabla c, \vec{u})$.

Expressions, in which tensors or tensor components are combined only by means of the tensor operations, are called the *tensor expressions*. They are convenient because they are valid with respect to an arbitrary basis.

4.2. SOME SPECIAL TENSORS

In addition to tensors of the general form, there are tensors with special properties.

1. **Zero tensor and the unit tensor.**

   Tensor, which maps each vector to the zero vector, is called the *zero tensor*. It will be denoted by O: $\vec{a} \mapsto 0$. It is easy to show that the matrix of the zero tensor with respect to arbitrary basis is the zero matrix.
   The identity mapping is called the *unit tensor*. It will be denoted by I : $\vec{a} \mapsto \vec{a}$. Also it is easy to show that the matrix of the unit tensor is always the identity matrix, and components of the unit tensor are $I_{ik} = \delta_{ik}$ *i.e.* are equal to the *Kronecker symbols*[2]. Thus, for each vector of a vector space, the equalities hold.

   $$O\vec{a} = 0, \qquad I\vec{a} = \vec{a}.$$

2. **Inverse tensor.**

   If A is a *one-to-one* tensor, then it is reversible, *i.e.* there exists a tensor A^{-1} called the *inverse tensor* with respect to A such that

   $$AA^{-1} = A^{-1}A = I.$$

   The zero tensor is an example of noninvertible (or degenerate) tensor. If tensors A and B are invertible, then their contraction AB is also invertible and

   $$(AB)^{-1} = B^{-1}A^{-1}.$$

   Besides,

   $$(A^{-1})^{-1} = A.$$

If $a \neq 0$ is a number, then

$$(a\mathrm{A})^{-1} = a^{-1}\mathrm{A}^{-1}.$$

3. **Transposed tensor.**

Let $\{e_i\}$ is an orthonormal basis, and $\vec{a} = a_i \vec{e}_i$, $\vec{b} = b_j \vec{e}_j$ and $\mathrm{A}\vec{a} = (\mathrm{A}_{ji}\, a_i)\, \vec{e}_j$. Consider the scalar product of the vectors $\mathrm{A}\vec{a}$ and $\vec{b}$:

$$(\mathrm{A}\vec{a}, \vec{b}) \overset{(1)}{=} (\mathrm{A}_{ji}a_i)b_j \overset{(2)}{=} a_i(\mathrm{A}_{ji}b_j) \overset{(3)}{=} a_i(\mathrm{A}_{ij}^{\mathrm{T}}b_j) = (\vec{a}, \mathrm{A}^{\mathrm{T}}\vec{b}). \qquad \textbf{(4.2)}$$

The action of a tensor on a vector in the component form is reduced to multiplication of a matrix by a column vector (this is the equality (1)). Under the existing agreement, the first index of the matrix element numbers the row and the second index numbers the column to which the element belongs. According to the rules of matrix multiplication, an element of the result is the sum of products of elements of the row of the matrix and corresponding elements of the column of the second factor. In order to interchange the order of summation in (4.2) (*i.e.* to use the equality (2)) and write all this in the matrix notation with preservation of the rules of matrix multiplication (as in the last expression), we need to introduce a new matrix A^{T} with the same components as that of the matrix A, but standing in another order: rows and columns are swapped (so we obtain the equality (3)). New matrix A^{T} is called the *transposed matrix* with respect to A. Accordingly, the tensor with matrix A^{T} with respect to the basis $\{\vec{e}_i\}$ is called the *transposed tensor* relative to A. Thus, one has (pay attention to the order of indices)

$$\mathrm{A}_{ij}^{\mathrm{T}} = \mathrm{A}_{ji}.$$

It is easy to verify the following properties of the operation of transposition:
$(\mathrm{A} + \mathrm{B})^{\mathrm{T}} = \mathrm{A}^{\mathrm{T}} + \mathrm{B}^{\mathrm{T}}$,
$(\mathrm{AB})^{\mathrm{T}} = \mathrm{B}^{\mathrm{T}}\mathrm{A}^{\mathrm{T}}$,
$(\mathrm{A}^{\mathrm{T}})^{\mathrm{T}} = \mathrm{A}$.
If A is an invertible tensor, then $(\mathrm{A}^{-1})^{\mathrm{T}} = (\mathrm{A}^{\mathrm{T}})^{-1}$.

4. **Symmetric and antisymmetric tensors.**

If

$$\mathrm{A} = \mathrm{A}^{\mathrm{T}},$$

the tensor A is called *symmetric*. The matrix of this tensor is symmetric with respect to the main diagonal, *i.e.*, $\mathrm{A}_{ij} = \mathrm{A}_{ji}$. Similarly, if

$$\mathrm{B} = -\mathrm{B}^{\mathrm{T}},$$

the tensor B is *antisymmetric*, and its matrix is antisymmetric with respect to the main diagonal, *i.e.*, $\mathrm{B}_{ij} = -\mathrm{B}_{ji}$.

The number of independent components of a matrix of the 2nd rank tensor is determined by its symmetry properties and dimensionality n of the corresponding vector space. So, if some tensor is of the general form, its matrix has n^2 independent components (each of its components does not dependent on others). The matrix of a symmetric tensor has $\frac{1}{2}n(n+1)$ independent components (the components on the main diagonal and in the upper or lower triangle). Finally, the matrix of an antisymmetric tensor has only $\frac{1}{2}n(n-1)$ independent components (these are components in upper or lower triangle, while the main diagonal contains zeros).

5. **Orthogonal tensor.**

Introduction in a vector space an additional structure, namely, the scalar product, allows one to give definition of the length of a vector $|\vec{a}| = \sqrt{(\vec{a}, \vec{a})}$. A tensor R is called *orthogonal* if it preserves the length of a vector, *i.e.*,

$$|R\vec{a}| = |\vec{a}|. \tag{4.3}$$

In this case, $R^{-1} = R^T$. Indeed, due to (4.2) and (4.3) the chain of equalities holds (here, numbers above the equal signs specify formulas, whereby the equalities hold)

$$|R\vec{a}|^2 = (R\vec{a}, R\vec{a}) \overset{(4.2)}{=} (\vec{a}, R^TR\vec{a}) \overset{(4.3)}{=} |\vec{a}|^2.$$

Inasmuch as $|\vec{a}|^2 = (\vec{a}, \vec{a})$ one obtains $R^T R = I$ and by virtue of definition of the inverse tensor $R^{-1} R = I$, we find $R^{-1} = R^T$.

6. **Positive definite tensor.**

Symmetric tensor A is called *positive definite* if for any non-zero vector $\vec{a}$ the inequality

$$(A\vec{a}, \vec{a}) > 0$$

holds. Geometrically, this means that the tensor A acting on a vector changes its direction by less than $\frac{\pi}{2}$ because the scalar product is positive only when the angle between vectors is less than $\frac{\pi}{2}$.

4.3. EIGENVALUE PROBLEM

Second-rank tensor transforms one vector into another. Consider the action of a reversible tensor A on an arbitrary non-zero vector $\vec{b}$. The result of this action is a new vector $A\vec{b}$, which generally does not coincide with the vector $\vec{b}$, *i.e.*, which has different length $|A\vec{b}| \neq |\vec{b}|$ and is rotated relative thereto by a certain angle.

Let the direction of the vector $\vec{b}$ changes smoothly, say, rotates clockwise. Since the vector $A\vec{b}$ is linearly dependent on the vector $\vec{b}$, its direction will also smoothly change. However, the speed of this change and its direction may be different. For this reason, after the rotation of the vector $\vec{b}$ at a certain angle directions of both vectors $\vec{b}$ and $A\vec{b}$ may coincide or be opposed. In both cases, the resulting vector $A\vec{b}$ is proportional to the initial vector $\vec{b}$. Mathematically, such a situation maybe written as the following equality

$$A\vec{b} = \lambda \vec{b}, \tag{4.4}$$

where λ is a number, and if this situation realizes, the vector $\vec{b}$ is called the *eigenvector* of the tensor A. The action of a tensor on its eigenvector is reduced, as we see, to stretching (compression) into $|\lambda|$ times and reflection, if $\lambda < 0$. The number λ is called the *eigenvalue* of the tensor A.

Eigenvalues and eigenvectors are important characteristics of tensors. To find them it is necessary to solve the problem (4.4), which in this case is called the *eigenvalue problem*. Each eigenvalue corresponds to an infinite number of eigenvectors. For instance, if $\vec{b}$ an eigenvector *i.e.* a vector satisfying the equation (4.4), any other vector $\alpha \vec{b}$, where α is arbitrary number, also satisfies this equation.

The eigenvalues of a positive definite tensor are positive. Indeed, if a tensor A is positive definite and a vector $\vec{b}$ is its eigenvector, *i.e.* $A\vec{b} = \lambda \vec{b}$, then

$$0 < (A\vec{b}, \vec{b}) = (\lambda \vec{b}, \vec{b}) = \lambda(\vec{b}, \vec{b}) = \lambda |\vec{b}|^2 \quad \Rightarrow \quad \lambda > 0.$$

NOTES

[1] Firstly, it is necessary to get used to it, as to any other new concept, and in general, to everything new. Habit gives rise if not to understanding, then at least, to its illusion and something habitual most often doesn't cause rejection.

[2] Kronecker Leopold (1823–1891), a German mathematician. Symbol bearing his name, is defined as

$$\delta_{ij} = \begin{cases} 0, i \neq j, \\ 1, i = j. \end{cases}$$

CHAPTER 5

Deformation

Abstract: We study an arbitrary deformation of a configuration of a body and show that this deformation may be reduced to a sequence of two special deformations: the rotation and stretching. The kinematics of a deformation is also considered. We introduce a notion of a vorticity vector and discuss its relationship with the antisymmetric part of the velocity gradient.

Keywords: Deformation, Deformation gradient, Deformation rate tensor, Kinematics of a deformation, Polar decomposition theorem, Rotation tensor, Stretching tensor, Velocity gradient, Vorticity vector.

5.1. STRETCHING AND ROTATION

During movement relative positions of points of a body, and, hence, its configuration are constantly changing. Points of the body move along their world-lines, each one with its own velocity. At the same time, by virtue of the continuity hypothesis, the movement of each point of a body is connected with movements of neighboring points. One way of investigation of this connection consists in a choice of a reference configuration κ and studying its deformation over time.

In general, at any given time the deformation of the reference configuration may be considered as a nonlinear coordinate transformation of points of a body $\mathbf{X} \mapsto \mathbf{x}(t)$. However, in the first approximation, in the vicinity of each point, it can be described by a linear dependence

$$dx = FdX, \tag{5.1}$$

where F is the tensor with components $F_{jk} = \partial_{X_k} x_j$.

Indeed, let $\mathbf{X}$ be some point in the reference configuration, and $\mathbf{X} + d\mathbf{X}$ be another point in the vicinity of $\mathbf{X}$ (see, Fig. **5.1**). As already mentioned, the quantities $\mathbf{X} = (X_1, X_2, X_3)$ and $\mathbf{X} + d\mathbf{X} = (X_1 + dX_1, X_2 + dX_2, X_3 + dX_3)$ are often called the radius-vectors of corresponding points of the space of places (see, p.40). In the current configuration these points are associated with the radius-vectors $\mathbf{x}(t, \mathbf{X})$ and $\mathbf{x}(t, \mathbf{X} + d\mathbf{X})$. To find out how the difference $d\mathbf{X}$ between points $\mathbf{X}$ and $\mathbf{X} + d\mathbf{x}$, has changed, we expand each component $\mathbf{x}(t, \mathbf{X} + d\mathbf{X})$ as a function of several variables in the Taylor series with respect to $\mathbf{x}(t, \mathbf{X})$.

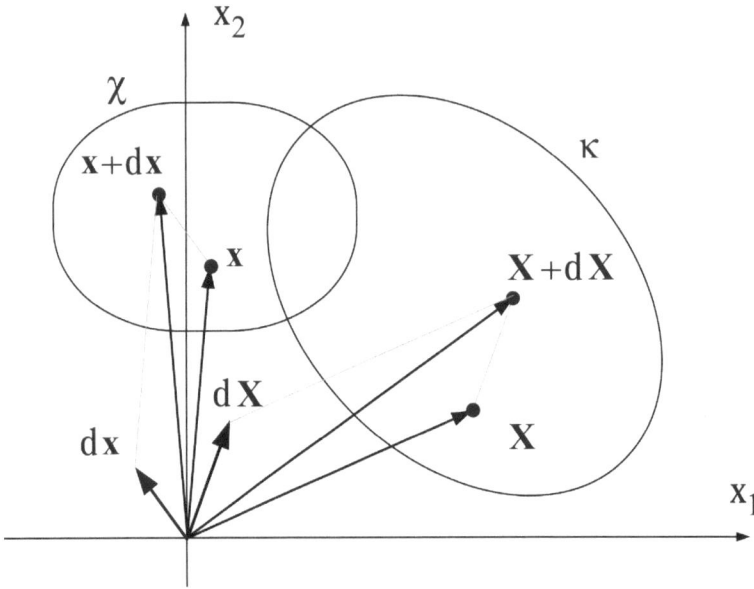

Fig. (5.1). Deformation of an increment of a radius-vector.

Limited to linear terms in the expansion we find (do not forget here the summation over k):

$$x_j(t, \mathbf{X} + d\mathbf{X}) = x_j(t, \mathbf{X}) + dX_k \, \partial_{X_k} x_j + o(|d\mathbf{X}|), \qquad \forall j = 1,2,3.$$

or what is the same

$$\mathbf{x}(t, \mathbf{X} + d\mathbf{X}) = \mathbf{x}(t, \mathbf{X}) + \mathbf{F} d\mathbf{X} + o(|d\mathbf{X}|).$$

The order symbol $o(\cdot)$ denotes terms of higher order of smallness than indicated in parentheses. In our case, these are terms, whose absolute values are small compared to $|d\mathbf{X}|$. Assuming now

$$dx \equiv x(t, X + dX) - x(t, X),$$

up to linear terms we obtain the expression (5.1).

The tensor F is called the *deformation gradient*. We had already encountered it when discussing the transfer theorem. The determinant J of the tensor F matrix

$$J = \begin{vmatrix} F_{11} & F_{12} & F_{13} \\ F_{21} & F_{22} & F_{23} \\ F_{31} & F_{32} & F_{33} \end{vmatrix} = \begin{vmatrix} \partial_1 x_1 & \partial_2 x_1 & \partial_3 x_1 \\ \partial_1 x_2 & \partial_2 x_2 & \partial_3 x_2 \\ \partial_1 x_3 & \partial_2 x_3 & \partial_3 x_3 \end{vmatrix}, \quad F_{jk} = \partial_k x_j \equiv \partial_{X_k} x_j, \quad \textbf{(5.2)}$$

is the Jacobian of the coordinate transformation of a point of a body in the reference configuration (the Lagrangian coordinates) into the coordinates of the same point of the body in the current configuration (the Euler coordinates).

5.1.1. Polar Decomposition Theorem

What is the deformation of the reference configuration (and therefore, of any other)? To answer this question, we use the *polar decomposition theorem*, according to which *any invertible tensor of rank 2 may be represented as a contraction of two tensors, one of which is orthogonal and the other is symmetric and positive definite.*

Since the theorem does not prescribe the order of tensors in this contraction, the deformation gradient F may be written down in two ways

$$F = RU = VR, \tag{5.3}$$

where R is an orthogonal tensor, while U and V are symmetric positive definite tensors (right and left depending on their position with respect to tensor R). If the basis is chosen, the expression (5.3) is understood as matrix equalities, and the contraction is regarded as a product of square matrices.

The action of the tensor F has a clear geometric meaning: the transformation of infinitesimal increment dX in the reference configuration into infinitesimal increment dx in the current configuration. Arbitrary deformation F, according to (5.3), may be interpreted as consecutive actions of two special deformations: first U and then R or, conversely, first R but then V *i.e.*

$$FdX = R(UdX) = V(RdX). \qquad\qquad (5.4)$$

It only remains to understand, what deformations are described by tensors R, U and V. We shall work using the orthonormal basis, with respect to which calculations are easier, and interpretation is more obvious.

5.1.2. Rotation Tensor

How tensor R acts on a vector? Any vector $\vec{a}$ from the given three-dimensional vector space is characterized by three numbers, its components with respect to, say, coordinate basis. In the case of spherical coordinates, these numbers are the lenght of the vector and two angles. The action of a tensor causes the change of these three numbers. The tensor R, as we have found out in the last chapter (p.61), leaves the length of any vector the same. Therefore, its action results in a change of direction or in the rotation of a vector. For this reason, the tensor R in this context, is called the *rotation tensor*.

5.1.3. Stretch Tensor

Let us now turn to tensors U and V .

1. Since both tensors U and V are real and symmetric, it is known that their eigenvalues are also real. Moreover, due to the polar decomposition it may be shown that the eigenvalues of the tensor U are equal to the eigenvalues of the tensor V. Indeed, let u_i be an eigenvalue of the tensor U and $\vec{\xi}_i$ be a corresponding eigenvector, and therefore the equality

$$U\vec{\xi}_i = u_i\vec{\xi}_i \qquad\qquad (5.5)$$

holds. Consider the vector $\vec{\xi}F_i$. Using the polar decomposition of the tensor F, we find

$$F\vec{\xi}_i = RU\vec{\xi}_i \overset{(5.5)}{=} Ru_i\vec{\xi}_i = u_i(R\vec{\xi}_i).$$

On the other hand,

$$F\vec{\xi}_i = VR\vec{\xi}_i.$$

Comparing both results, we obtain

$$V(R\vec{\xi}_i) = u_i(R\vec{\xi}_i)$$

or, denoting $\vec{\zeta}_i = R\vec{\xi}_i$,

$$V\vec{\zeta}_i = u_i\vec{\zeta}_i.$$

This equality means that $\vec{\zeta}_i$ is an eigenvector of the tensor V whereas u_i is its eigenvalue. Thus, not only do we have shown that u_i is the eigenvalue of the tensor V but also have found a relationship between eigenvectors of both tensors U and V. If $\vec{\xi}_i$ is an eigenvector of the tensor U corresponding to the eigenvalue u_i, an eigenvector of the tensor V corresponding to the same eigenvalue, is the vector $\vec{\zeta}_i = R\vec{\xi}_i$ which have the same length as $\vec{\xi}_i$ but another direction.

2. Regarding eigenvectors of a tensor the following assertion is true: eigenvectors corresponding to different eigenvalues of a real symmetric tensor are orthogonal, *i.e.*, $(\vec{\xi}_i, \vec{\xi}_j) = 0$ when $i \neq j$. Indeed, let $\vec{\xi}_i$ and $\vec{\xi}_j$ are eigenvectors of the tensor U and u_i and u_j are corresponding eigenvalues, and $u_i \neq u_j$. The following chain of equalities holds:

$$u_i(\vec{\xi}_i, \vec{\xi}_j) = (u_i\vec{\xi}_i, \vec{\xi}_j) \overset{(5.5)}{=} (U\vec{\xi}_i, \vec{\xi}_j) \overset{(4.2)}{=} (\vec{\xi}_i, U^T\vec{\xi}_j)$$

Further, due to the symmetry of the tensor U, one has

$$u_i(\vec{\xi}_i, \vec{\xi}_j) = (\vec{\xi}_i, U\vec{\xi}_j) \overset{(5.5)}{=} (\vec{\xi}_i, u_j\vec{\xi}_j) = u_j(\vec{\xi}_i, \vec{\xi}_j).$$

Now subtracting the last expression from the first one, we obtain

$$0 = (u_i - u_j)(\vec{\xi}_i, \vec{\xi}_j).$$

Thus, if $u_i \neq u_j$ by necessity $(\vec{\xi}_i, \vec{\xi}_j) = 0$, which means the orthogonality of $\vec{\xi}_i$ and $\vec{\xi}_j$. If the tensors U and V have three distinct eigenvalues, then they also have three orthogonal eigenvectors, and these three vectors are related by

$$\vec{\zeta}_i = R\vec{\xi}_i,$$

where $\vec{\xi}_i$ is an eigenvector of the tensor U and $\vec{\zeta}_i$ is an eigenvector of the tensor V; both eigenvectors correspond to the eigenvalue u_i. These eigenvectors define directions, which are called *principal axes of strain*. The tensor U defines them in the reference configuration κ *i.e.* at some point $\mathbf{X}$ and the tensor V defines them in the current configuration χ *i.e.* at the corresponding point $\mathbf{x}(t, \mathbf{X})$. Rotation tensor R transforms the principal axes of strain at the point $\mathbf{X}$ into the principal axes of strain at the point $\mathbf{x}$.

3. Among other things, the polar decomposition theorem states that both tensors U and V are positive definite, *i.e.*, those for which the inequality

$$(U\vec{a}, \vec{a}) > 0 \quad \forall \vec{a} \neq 0$$

is valid.

If a tensor, say U , *is positive definite*, then as it was shown on p.62 *its eigenvalues are positive*. This follows from the fact that if u_i is an eigenvalue, and $\vec{\xi}_i$ is a corresponding eigenvector of the tensor U then one has

$$(U\vec{\xi}_i, \vec{\xi}_i) = u_i(\vec{\xi}_i, \vec{\xi}_i) = u_i|\vec{\xi}_i|^2 > 0 \quad \Rightarrow \quad u_i > 0.$$

The positivity of eigenvalues allows their interpretation as ratios of lengths of vectors $F\vec{\xi}_i$ (images of vectors $\vec{\xi}_i$) in the current configuration to lengths of vectors $\vec{\xi}_i$ in the reference configuration and call them *main stretchings*.

The tensors U and V thus describe a stretch (or compression) in three orthogonal directions. The magnitudes of stretching are determined by eigenvalues, and directions in which they occur are determined by corresponding eigenvectors. The tensor U (or V) is called the right (left) *stretch tensor*.

If R = I *i.e.* the rotation is absent, then F = U = V and deformation is called *pure stretching*. If, on the contrary, U = V = I, then F = R, there is no stretching and deformation is called *rotation*.

Thus, the polar decomposition theorem states that deformation, which is locally described by F may be obtained in two ways: either by accomplishing a pure stretching along three mutually orthogonal directions $\vec{\xi}_i$ with main stretchings u_i and subsequent rotation of these directions, or, first, performing rotation and then, the same stretching u_i but along new directions $\vec{\zeta}_i = R\vec{\xi}_i$.

5.2. KINEMATICS OF DEFORMATION

In order to understand how deformation proceeds, let's calculate the time derivative of (5.1) at point t_0 and obtain

$$d_t(d\mathbf{x}) = (d_t F)d\mathbf{X}. \tag{5.6}$$

The left side is the increment of velocity $d_t(d\mathbf{x}) = d(d_t\mathbf{x}) = d\vec{v}$ in the vicinity of a given point. A component of the tensor $d_t F$ in the right side is equal to

$$(d_t F)_{ki} = d_t(\partial_{X_i} x_k) = \partial_{X_i}(d_t x_k) = \partial_{X_i} v_k. \tag{5.7}$$

The last equality holds because the Lagrangian variables do not depend on time. Components of a vector, tangent to the trajectory of a point of a body are functions of time, and current (Euler) coordinates of the point. Due to relation (2.8)

$$v_k(t, \mathbf{x}) = v_k(t, \mathbf{x}(t, \mathbf{X}))$$

the velocity components turn out to be composite functions of the Lagrangian coordinates of the body. For this reason, the chain of equalities (5.7) may be extended by differentiating the velocity components as follows

$$(d_t F)_{ki} = \left(\partial_{x_j} v_k\right) \partial_{X_i} x_j = (\nabla \vec{v})_{kj} F_{ji},$$

or in the tensor notation

$$d_t F = GF, \tag{5.8}$$

where $G \equiv \nabla \vec{v}$. This gives us the expression for the velocity gradient G in terms of the deformation gradient F:

$$G = (d_t F) F^{-1}. \tag{5.9}$$

Note, that the inverse tensor F^{-1} is placed here to the right of time derivative (and not to the left), because, trying to obtain the expression for G , we compute the contraction of both sides of (5.8) with F^{-1} , and tensors in contraction do not commute. Thus, the velocity increment is equal to

$$d\vec{v} = (d_t F) d\mathbf{X} = (GF) d\mathbf{X} = G(F d\mathbf{X}) = G d\mathbf{x} \tag{5.10}$$

Given the representation (5.3), we obtain

$$G = ((d_t R)U + R d_t U) F^{-1} = ((d_t V)R + V d_t R) F^{-1}. \tag{5.11}$$

Let's try to find a suitable interpretation of these expressions.

Any tensor and, in particular, G can be written as a sum of symmetric D and antisymmetric W parts:

$$G = D + W = \underbrace{\tfrac{1}{2}(G + G^T)}_{D} + \underbrace{\tfrac{1}{2}(G - G^T)}_{W}. \tag{5.12}$$

Symmetric tensor D

$$D_{ik} = \tfrac{1}{2}(\nabla_k v_i + \nabla_i v_k) \tag{5.13}$$

is called the *deformation rate tensor*, and antisymmetric tensor W

$$W_{ik} = \tfrac{1}{2}(\nabla_k v_i - \nabla_i v_k) \tag{5.14}$$

we shall call *spin*[1].

At time t_0 the deformation gradient is equal to the unit tensor (deformation is absent), where

$$R = U = V = I \quad \Rightarrow \quad F = F^{-1} = I. \tag{5.15}$$

If we assume that in the vicinity of t_0 the deformation gradient is nearly the unit tensor (*i.e.*, the equalities (5.15) in a first approximation are valid), then (5.11) gives

$$G \approx d_t R + d_t U \approx d_t V + d_t R. \tag{5.16}$$

Hence

$$D \approx d_t U \approx d_t V, \qquad W \approx d_t R. \tag{5.17}$$

Due to its symmetry, the tensor D has three real eigenvalues (*principal rates of deformation*) and three orthogonal eigenvectors (*principal axes of deformation rate tensor*).

Thus, as the first approximation, the velocity increment consists of two parts: symmetric part, connected with the action of the tensor D, and antisymmetric part, associated with the action of the tensor W:

$$d\vec{v} = D d\mathbf{x} + W d\mathbf{x}. \tag{5.18}$$

The tensor D describes the rate of expansion/compression in three orthogonal directions, and the antisymmetric tensor W describes the rate of rotation of these directions.

Since in the three-dimensional case, which is regarded here, the number of independent components of the tensor W is equal to the number of components of a vector (see, p. 61), instead of the tensor W a corresponding vector may be considered[2].

Indeed, let's consider the quantity W$d\mathbf{x}$:

$$
\begin{aligned}
W d\mathbf{x} &= \begin{pmatrix} 0 & W_{12} & W_{13} \\ W_{21} & 0 & W_{23} \\ W_{31} & W_{32} & 0 \end{pmatrix} \begin{pmatrix} dx_1 \\ dx_2 \\ dx_3 \end{pmatrix} \\
&= \begin{pmatrix} 0 & -W_{21} & W_{13} \\ W_{21} & 0 & -W_{32} \\ -W_{13} & W_{32} & 0 \end{pmatrix} \begin{pmatrix} dx_1 \\ dx_2 \\ dx_3 \end{pmatrix}
\end{aligned}
$$

and introduce the following notation:

$$
\begin{aligned}
\omega_1 &\equiv 2W_{32} = \partial_{x_2} v_3 - \partial_{x_3} v_2, \\
\omega_2 &\equiv 2W_{13} = \partial_{x_3} v_1 - \partial_{x_1} v_3, \\
\omega_3 &\equiv 2W_{21} = \partial_{x_1} v_2 - \partial_{x_2} v_1.
\end{aligned}
$$

If the numbers ω_1, ω_2 and ω_3 are considered as components of the vector $\vec{\omega}$, then $\vec{\omega} = \nabla \times \vec{v}$. This follows from the definition of the vector product. The vector $\vec{\omega}$ is called the *vorticity vector*. It determines the angular velocity of a fluid at a point. Substituting the components of the vorticity vector in the expression for W$d\mathbf{x}$, we find:

$$W d\mathbf{x} = \frac{1}{2} \begin{pmatrix} 0 & -\omega_3 & \omega_2 \\ \omega_3 & 0 & -\omega_1 \\ -\omega_2 & \omega_1 & 0 \end{pmatrix} \begin{pmatrix} dx_1 \\ dx_2 \\ dx_3 \end{pmatrix} =$$

$$= \frac{1}{2} \begin{pmatrix} -\omega_3 dx_2 + \omega_2 dx_3 \\ \omega_3 dx_1 - \omega_1 dx_3 \\ -\omega_2 dx_1 + \omega_1 dx_2 \end{pmatrix}.$$

The resulting vector is nothing but the vector product

$$W d\mathbf{x} = \frac{1}{2} \vec{\omega} \times d\mathbf{x}. \tag{5.19}$$

Now using the expression for the infinitesimal velocity increment $d\vec{v} = \vec{v}(t, \mathbf{x} + d\mathbf{x}) - \vec{v}(t, \mathbf{x})$ we find the velocity in the vicinity of $\mathbf{x}$:

$$\vec{v}(t, \mathbf{x} + d\mathbf{x}) \overset{(5.18)}{=} \vec{v}(t, \mathbf{x}) + D d\mathbf{x} + W d\mathbf{x} = \vec{v}(t, \mathbf{x}) + D d\mathbf{x} + \frac{1}{2} \vec{\omega} \times d\mathbf{x}. \tag{5.20}$$

The interpretation of this decomposition is as follows: the instantaneous state of a medium in the vicinity $d\mathbf{x}$ of the considered point $\mathbf{x}$ consists of the translational motion with velocity $\vec{v}(t, \mathbf{x})$ stretching/compression in three orthogonal directions $(D d\mathbf{x})$ and rotation of these directions $(W d\mathbf{x} = \frac{1}{2} \vec{\omega} \times d\mathbf{x})$[3]. If $D = 0$ then stretching does not occur and the fluid moves as a rigid body. If $W = 0$, the stretching directions are not changed and the motion is called irrotational. Conversely, if $W \neq 0$, the motion is rotational.

NOTES

[1] In the hydrodynamic literature, this tensor usually remains nameless, probably because its function is successfully discharged by the vorticity vector $\vec{\omega}$ (see below). Here this tensor is called spin after C.Trusdell [1].

[2] More precisely, a pseudo, or axial vector, *i.e.*, an object which is sensitive to a change of a basis: the reflection of the basis causes the change of the sign of a pseudo vector.

[3] Note, how the continuity hypothesis is working here. Having information on the motion of continuum ($\vec{v}$, D, W or $\vec{\omega}$) only in one point $\mathbf{x}$ we, nevertheless, due to

the accepted hypothesis, are able to make judgments about the motion of the medium at any point in a vicinity of the given one. Of course, the further from the point **x** the more erroneously this judgment is, but the estimation of this error is known (of order of magnitude $|d\mathbf{x}|^2$).

CHAPTER 6

The Continuity Equation

Abstract: Starting from this chapter and to the end of the first part we are building a closed system of equations of a fluid model. Firstly, we discuss differential conservation laws, and related balance equations. All previously obtained expressions are applied then to the mass of a body, and this gives the first equation of the fluid model, the continuity equation.

Keywords: Balance equation, Compressibility, Continuity equation, Differential conservation law, Flow through a surface, Flux, Levi-Civita symbol, Mass flux density, Rate of volume expansion.

6.1. THE RATE OF VOLUME EXPANSION

We have already noted that the value of the Jacobian J is equal to relative expansion or compression of an infinitesimal volume of a moving body. The rate of change of this quantity is related to the *rate of volume expansion*, which is defined as the ratio $\frac{1}{J} d_t J$. The Jacobian J of transformation of the Lagrangian coordinates into Euler coordinates is the determinant of the deformation gradient F (see. p.65).

Every determinant of a matrix is a sum, any term of which is the product that contains one and only one element from each row and each column of the matrix. The number of terms in the sum is equal to the number of variants of such products. Thus, any determinant does not include products which do not contain elements of some row/column or contain them more than one.

When working with determinants it is convenient to use the so-called *Levi-Civita symbols*[1]. These antisymmetric[2] symbols allow one to make expression of a determinant compact and automatically to trace signs before terms. They are defined as follows:

$$\varepsilon^{ijk} = \begin{cases} 1, & \text{even permutation of indices}, \\ 0, & \text{coinciding indices}, \\ -1, & \text{odd permutation of indices}. \end{cases}$$

Even/odd permutation means the number of permutations of neighboring indices of the given combination, which needs to be done to obtain a standard combination, and the combination 123 is regarded as such. It may be shown that although a lot of ways lead to the standard combination, all of them are either even, or odd, and for each combination such option is always one.

Let A be a square 3×3 matrix with elements a_{ij} , then its determinant may be written either as

$$\det(A) = \varepsilon^{ijk} a_{i1} a_{j2} a_{k3}, \tag{6.1}$$

or as

$$\det(A) = \varepsilon^{ijk} a_{1i} a_{2j} a_{3k}.$$

The difference in these expressions is in order of indices of elements a_{ij}. In both cases any term in these triple sums contains one and only one element of each row and each column of the matrix A. In the first case symbols ε^{ijk} select factors by columns (for example, a_{i1} is the i-th element of the 1-st column) and by rows in the second case. Of all possible combinations of $a_{i1} a_{j2} a_{k3}$ or $a_{1i} a_{2j} a_{3k}$ the Levi-Civita symbol eliminates superfluous.

If explicitly write out any of three sums in (6.1), for example, for i, we obtain the expansion of determinant by elements of the column (the 1st column in our case):

$$\det(A) = a_{11}(\varepsilon^{1jk} a_{j2} a_{k3}) + a_{21}(\varepsilon^{2jk} a_{j2} a_{k3}) + a_{31}(\varepsilon^{3jk} a_{j2} a_{k3}). \tag{6.2}$$

Terms in brackets $A^{1i} \equiv \varepsilon^{ijk} a_{j2} a_{k3}$ are the so-called cofactors of the matrix elements a_{i1}. Thus,

$$\det(A) = a_{11}A^{11} + a_{21}A^{12} + a_{31}A^{13} = a_{i1}A^{1i}. \tag{6.3}$$

Since any determinant may be expanded by elements of each column (not necessarily the first one), we have

$$\det(A) = a_{ij}A^{ji}, \quad j = 1,2,3. \tag{6.4}$$

Here the summation is implied over the index i only and the index j may take any value, as specifically indicated. It is also possible to expand a determinant by elements of a row, *i.e.* one may write:

$$\det(A) = a_{ij}A^{ji}, \quad i = 1,2,3. \tag{6.5}$$

This is the expansion by elements of the i-th row (the summation over j is implied). Both expressions (6.4) and (6.5) may be written in a more general form

$$a_{kj}A^{ji} = a_{jk}A^{ij} = \delta_{ki}\det(A), \tag{6.6}$$

if we use the well-known property of cofactors: the sum of products of elements of one row (column) of the matrix and cofactors of elements of another row (column) is equal to zero.

Let us, now, calculate the time derivative of Jacobian J, which in accordance with (6.1) may be written as

$$J = \det(F) = \varepsilon^{ijk}F_{i1}F_{j2}F_{k3} = \varepsilon^{ijk}\,\partial_1 x_i\,\partial_2 x_j\,\partial_3 x_k.$$

Here we again use the abbreviation $\partial_i x_k \equiv \partial_{X_i} x_k$. Differentiating J with respect to time, we obtain

$$d_t J = d_t\left(\varepsilon^{ijk}\,\partial_1 x_i\,\partial_2 x_j\,\partial_3 x_k\right)$$
$$= \varepsilon^{ijk}\left(d_t(\partial_1 x_i)\,\partial_2 x_j\,\partial_3 x_k + \partial_1 x_i d_t(\partial_2 x_j)\,\partial_3 x_k + \partial_1 x_i\,\partial_2 x_j d_t(\partial_3 x_k)\right).$$

Recall that the Lagrangian variables are independent on time and, therefore, differentiating with respect to Lagrangian variable and with respect to time may trade places $d_t(\partial_m x_i) = \partial_m(d_t x_i) = \partial_m v_i$ and it is possible to write:

$$d_t J = \varepsilon^{ijk}\left(\partial_1 v_i\,\partial_2 x_j\,\partial_3 x_k + \partial_1 x_i\,\partial_2 v_j\,\partial_3 x_k + \partial_1 x_i\,\partial_2 x_j\,\partial_3 v_k\right).$$

We denote by J^{ki} the cofactor of ki-th element of the Jacobian matrix (deformation gradient F). Introduction of cofactors allows simplifying the expression:

$$d_t J = (\partial_m v_i)J^{mi}.$$

Now we take into account that components of velocities are given as functions of the Euler variables $v_i = v_i(t, \mathbf{x}(t))$ Using a one-to-one correspondence between the Euler and Lagrangian variables (see. p. 27), we write $v_i = v_i(t, \mathbf{x}(t, \mathbf{X}))$. Now differentiating the velocity components v_i as composite functions, we find $\partial_m v_i = \partial_{x_k} v_i \partial_m x_k$. Next, we use the formula (6.6) and obtain

$$d_t J = \left(\partial_{x_k} v_i \, \partial_m x_k \right) J^{mi} = \partial_{x_k} v_i \left(\partial_m x_k J^{mi} \right)$$
$$= \partial_{x_k} v_i \delta_{ki} J = J \partial_{x_k} v_k.$$

The latter expression is the product of the Jacobian and divergence of velocity. So we get

$$d_t J = J(\nabla, \vec{v}) = J \, div \vec{v}. \qquad (6.7)$$

Thus, the relative rate of a volume expansion (or compression) is determined by the velocity field $\vec{v}$ and is equal to its divergence

$$\frac{1}{J} d_t J = (\nabla, \vec{v}). \qquad (6.8)$$

6.2. DIFFERENTIAL CONSERVATION LAWS AND BALANCE EQUATIONS

Let's continue consideration of differential conservation laws. In the second chapter, we found out that if an integral parameter $\Pi(t, \mathcal{B})$ satisfies the integral conservation law (formula 2.10)

$$d_t \Pi = \Sigma,$$

its density π satisfies the corresponding differential conservation law (2.19):

$$d_t (\pi J) = \sigma J, \qquad (6.9)$$

or (2.20)

$$d_t \pi + \frac{\pi}{J} d_t J = \sigma.$$

Taking into account (6.7), the latter expression may be written in the form

$$d_t \pi + \pi (\nabla, \vec{v}) = \sigma, \tag{6.10}$$

which is also known as the *differential conservation law*.

Further we write the total derivative $d_t \pi$ using the Euler's formula and take into account that

$$(\vec{v}, \nabla \pi) + \pi (\nabla, \vec{v}) = (\nabla \pi, \vec{v}) + \pi (\nabla, \vec{v}) = (\nabla, \pi \vec{v}).$$

All this gives us another variant of the equation (6.9)

$$\partial_t \pi + (\nabla, \pi \vec{v}) = \sigma, \tag{6.11}$$

which is called the *differential balance equation* of the density of parameter Π. The quantity $\pi \vec{v}$ is called the *flux density* of the parameter Π.

Obviously, (6.9) is the same as (6.11), but has different emphasis. In the first case it is argued that the rate of change of the quantity πJ in those places which a point of the body passes, while moving along the world-line, is determined by the value of σJ, *i.e.* it is proportional to its generation σ. In the second case it is indicated that change in the density π at a given location is generated by a combined effect of generation and the flux density $\pi \vec{v}$. Note that both equations are valid for the density of any conserved integral parameter.

Since the flux density of a parameter exists, corresponding integral parameter (*i.e.* the *flux of a parameter* Π) also should exist. Indeed, such quantity may be defined. It naturally appears when an integral parameter $\Pi(t, V)$ characterizing a certain spatial domain is considered. By definition (see. p.25) it is equal to

$$\Pi(t, V) = \int_V \pi \, dV.$$

How does this quantity change in time? Unlike the parameter $\Pi(t, \mathcal{B})$, which is a function of t only, the parameter $\Pi(t, V)$ is also a differentiable function of the volume V. Thus, the rate of change of $\Pi(t, V)$ is determined by the partial time derivative $\partial_t \Pi(t, V)$ computed at a fixed argument V. Since V is fixed, the differentiation and integration may be interchanged, and we get the following equation:

$$\partial_t \Pi(t, V) = \int_V \partial_t \pi \, dV. \tag{6.12}$$

Using the formula (6.11) we express the derivative of the density:

$$\partial_t \pi = \sigma - (\nabla, \pi \vec{v}), \tag{6.13}$$

and obtain

$$\partial_t \Pi(t, V) = \int_V \sigma \, dV - \int_V (\nabla, \pi \vec{v}) \, dV. \tag{6.14}$$

The first term on the right hand side of this equation is called the *production* of the parameter Π in the *volume V*, while the second term is called the *inflow* of the quantity Π into the *volume V*.

Applying the Gauss' theorem[3] to the last integral in (6.14) and taking into account that the integral $\int_V \sigma \, dV$ is equal to Σ, the generation of the parameter Π (see the definition in (2.12)), we find

$$\partial_t \Pi(t, V) = \Sigma - \int_S (\pi \vec{v}, \vec{n}) \, dS. \tag{6.15}$$

Here S is the surface bounding the fixed volume V and $(\pi\vec{v}, \vec{n}) = (\pi\vec{v})_n$ is the projection of the flux density of Π in the direction of the outer normal $\vec{n}$ to S. The second term in the right-hand side of the equation (6.15) is called the *flow* of parameter Π *through the surface S*. If the integral is positive, the resulting flux through the surface S is directed outwards.

Both equations (6.14) and (6.15) describe the balance of parameter Π in V and therefore called the *balance equations* of Π. The interpretation of these equations is obvious: the change of parameter Π in the region V is caused (balanced) by the production inside the region V (*i.e.*, there are sources/sinks with power Σ of the quantity Π inside V) and the influx from the outside (just so is in the equation (6.14)) or, equivalently, by the flow of parameter Π across the border S of the region V (and so is in the equation (6.15)). The rate of change of Π is equal to the sum of production and influx. These equations as well as the differential conservation law (2.10) are valid for any conserved integral parameter.

6.3. THE CONTINUITY EQUATION

Now we apply all that has been said, to the already known integral parameter, the mass of the body $M(\mathcal{B})$ with density $\rho(t, \mathbf{x}(t))$ and zero production (see. p.12). Using the equations (2.11), (2.19), (6.11) and (6.15) we immediately obtain the following relations.

1. The equation (2.11) gives the integral *law of conservation of mass*

$$d_t M(\mathcal{B}) = 0. \tag{6.16}$$

2. The equation (2.19) gives the differential law of conservation of mass

$$\frac{1}{J} d_t(\rho J) = 0, \tag{6.17}$$

and the expression (6.11) gives another, more often used form of the same differential mass conservation law, called the *continuity equation*:

$$d_t\rho + \rho(\nabla, \vec{v}) = 0 \quad \text{or} \quad \partial_t\rho + (\nabla, \rho\vec{v}) = 0. \tag{6.18}$$

Here the quantity $\rho\vec{v}$ is called the *mass flux density*. Resolving the first equation in (6.18) with respect to divergence of velocity, we find

$$(\nabla, \vec{v}) = -\frac{1}{\rho} d_t \rho, \tag{6.19}$$

which corresponds to (2.20) and means that the divergence of velocity is equal to the relative *compressibility* of the medium.

If the mass density of a fluid does not change when moving along the world-lines, then there is no change in volume and the divergence of velocity is absent

$$d_t \rho = 0 \quad \Leftrightarrow \quad d_t J = 0 \quad \Leftrightarrow \quad (\nabla, \vec{v}) = 0. \tag{6.20}$$

Such medium is called *incompressible*, and each of the equations (6.20) is called the *continuity equation of incompressible fluid*.

On the contrary, if $d_t \rho \neq 0$ the medium is called *compressible*. In the region, where the velocity field has positive values of divergence, the values of the mass density decrease (the rate of change of the mass density is negative). Conversely, when the velocity field has negative values of divergence, the values of the mass density increase.

3.   The *mass balance equation in a fixed volume* V

$$\partial_t M(t, V) = -\int_S (\rho \vec{v}, \vec{n}) dS \tag{6.21}$$

states that the rate of change of the mass in the volume V (left-hand side of the equation) is equal to the *mass flux across the boundary* S (right-hand side of the equation).

The continuity equation (or its equivalent) is the first but not last equation of the model of a moving continuum. It contains two unknown functions ρ and $\vec{v}$ and therefore is underdetermined. If we are given a velocity field, the equation (6.18) allows one to determine the mass density field and to trace its evolution. Otherwise, an infinite number of solutions exists. Thus, to find both fields (the mass density and velocity) another vector equation is required. We are going to derive it in the next chapter.

NOTES

[1] Levi-Civita Tulio (1873–1941), an Italian mathematician.

[2] As in the case of tensors of rank 2 antisymmetry manifests itself in the change of sign under permutation of neighboring indices.

[3] Gauss Karl Friedrich (1777–1855), a German mathematician.

CHAPTER 7

Fluid Dynamics

Abstract: We postulate the basic principle of dynamics and derive the second equation of the fluid model, the equation of motion. The dynamics of the fluid is investigated. We introduce the notion of momentum of the body, study the rate of change of momentum and discuss forces acting on a body. The body forces and the contact forces and corresponding densities are considered. The equation of motion in the Cauchy form is obtained. Finally, we propose the simplest form of the stress tensor which gives the equation of motion of the perfect fluid.

Keywords: Basic principle of dynamics, Body force, Center of mass, Contact force, Equation of motion, Momentum, Momentum balance equation, Normal stress, Orientation, Perfect fluid, Potential, Shear stress, Stress tensor.

7.1. MOMENTUM OF A BODY AND THE RATE OF ITS CHANGE. THE BASIC PRINCIPLE OF DYNAMICS

We have already considered forces as a means of describing interactions of bodies, and have also defined the structure of the vector space on the system of forces defined on the set Ω. We link now these representations with spatial-temporal motion pattern. We introduce the notion of the *center of mass of a configuration of a body* $\mathcal{B}$ and define it *via* the expression

$$\mathbf{x}_M(t) = \frac{1}{M(\mathcal{B})} \int_\chi \rho \mathbf{x} dV. \tag{7.1}$$

The function $\mathbf{x}_M(t)$ specifies the trajectory, along which would move the point with mass equal to the mass of the body $M(\mathcal{B})$. Vector $\vec{v}_M = d_t \mathbf{x}_M(t)$ tangent to this trajetory, is the velocity of the center of mass of the body.

A moving body may be associated with vector $\mathbf{m}$,

$$\mathbf{m}(t, \mathcal{B}) = \vec{v}_M(t) M(\mathcal{B}), \tag{7.2}$$

which characterizes both its mass and its velocity. This characteristic is called the *momentum* of the body $\mathcal{B}$. The following statement makes up the basic principle of dynamics.

The rate of change of momentum of a body is equal to the force acting on it:

$$d_t \mathbf{m}(t, \mathcal{B}) = \mathbf{f}_{\mathcal{B}}(t). \tag{7.3}$$

This is a postulate, *i.e.* a connection between physical characteristics of a moving body, which is derived from experience and does not require the proof. In this form it holds in inertial frames of reference (see footnote 12 p.31) and is the foundation of mechanics in general (the Newton's equations of motion of a system of material points) and fluid mechanics, in particular.

The momentum is an integral characteristic of a body. In order to calculate the density of momentum, we shall derive a useful formula.

Let $h = h(t, \mathbf{x}(t))$ is an arbitrary differentiable function, and ρ is the mass density (*i.e.* a function satisfying the equation (6.18)). Then, using the transfer theorem (2.18) and the continuity equation (6.17), we write down the following chain of equalities:

$$d_t \int_\chi \rho h \, dV \overset{(2.18)}{=} \int_\chi \frac{1}{J} d_t(\rho h J) \, dV = \int_\chi (\rho d_t h + \frac{h}{J} \underbrace{d_t(\rho J)}_{=0}) \, dV,$$

and finally obtain

$$d_t \int_\chi \rho h \, dV = \int_\chi \rho d_t h \, dV. \tag{7.4}$$

This is the required formula which holds for arbitrary smooth function h.

Now we substitute (7.1) in (7.2) and by virtue of (7.4) get:

$$\mathbf{m}(t, \mathcal{B}) = d_t \int_\chi \rho \mathbf{x} \, dV = \int_\chi \rho \vec{v} \, dV. \tag{7.5}$$

In accordance with the definition (2.3), we find that the density of momentum is equal to $\rho\vec{v}$ *i.e.,* to the mass flux density. The basic principle of dynamics may be written now as follows

$$d_t\mathbf{m}(t,\mathcal{B}) = \int_\chi \rho d_t\vec{v}dV = \mathbf{f}_\mathcal{B}(t). \tag{7.6}$$

Here we again have used the formula (7.4).

The basic principle in the form of (7.3) or (7.6) is the equality between integral quantities. In order to obtain corresponding differential equation we proceed as described in the section 2.4.2. But first it is necessary to define the density of the force and write the force in the right side of (7.6) as an integral over the current configuration of the body. To this we now turn.

7.2. BODY FORCES AND CONTACT FORCES

Discussing forces, we have also introduced the notion of the *resultant force* $\mathbf{f}=\mathbf{f}_\mathcal{B}(\mathcal{B}^e,t)$ acting on the body $\mathcal{B}$ by its exterior $\mathcal{B}^e$ *i.e.* the sum of all forces acting on the body. At present, resultant force is usually regarded as a sum of two vectors, which reflect different nature of acting forces.

The thing is that only a part of points of $\mathcal{B}^e$, which forms a subset $\mathcal{B}^e_s$ is in direct contact with the body $\mathcal{B}$. Their interaction with points of the body is called contact interaction and the force, describing this interaction, is called the contact force $\mathbf{f}_c=\mathbf{f}_\mathcal{B}(\mathcal{B}^e_s,t)$. The remaining points of $\mathcal{B}^e$ included in the set $\mathcal{B}^e_o$ interact with the body $\mathcal{B}$ indirectly by means of a field (or by radiation). The force describing such interaction is called the body force $\mathbf{f}_B=\mathbf{f}_\mathcal{B}(\mathcal{B}^e_o,t)$. Thus , we represent the exterior of the body as a union

$$\mathcal{B}^e = \mathcal{B}^e_s \cup \mathcal{B}^e_o, \tag{7.7}$$

and the force of interaction as corresponding sum

$$\mathbf{f} = \mathbf{f}_c + \mathbf{f}_B. \tag{7.8}$$

Such a representation of forces is quite justified because any kinematic boundary consists of the same points of the body and, hence, the sets $\mathcal{B}^e_s$ and $\mathcal{B}^e_o$ do not change in time.

Thus, the equation (7.3) is now written in the form

$$d_t \mathbf{m} = \mathbf{f}_C + \mathbf{f}_B. \tag{7.9}$$

Since both components of the force $\mathbf{f}$ are integral parameters of the body (just like the momentum), we write them by analogy with (7.5) as integrals over the appropriate domains (the volume of the current configuration and the surface which bounds this volume) of the respective densities. Let us assume by definition

$$\mathbf{f}_C = \int_{\partial \chi} \vec{\tau} dS, \tag{7.10}$$

$$\mathbf{f}_B = \int_{\chi} \rho \vec{b} dV. \tag{7.11}$$

Here $\rho \vec{b}$ is the *body force density*, a function defined on points of the body, and $\vec{\tau}$ is the *contact force density*, commonly referred to as *stress*, a function defined on points of the body, which form the surface of configuration $\partial \chi$ (kinematic boundary) and dS is the infinitesimal element of the surface.

It should be noted that the representations (7.7) and (7.8) are valid for an arbitrarily small part of the body with its own exterior and, therefore, its own balance (7.9). The contact force $\mathbf{f}_C$ is defined on the surface of every conceivable part of the body, and the stress field $\vec{\tau}$ is defined at all points of configuration of the body. However, at points, within the body in question, all stresses are mutually compensated, and uncompensated remain stresses only at the external boundary of the body. In such form (7.10) these external contact forces are included into the balance (7.9). If the stress field is known, the resultant contact force is defined and does not depend on what may occur at points not lying on the boundary (*i.e.*, surface) $\partial \chi$.

The specific density of the body force $\vec{b}(t, \mathbf{x}(t))$ is a given function of time and place, which does not depend on a particular body $\mathcal{B}$. Therefore, the force $\mathbf{f}_B$ is called the *external body force*. Time-independent field $\vec{b}$ is called *stationary*. If the density $\vec{b}$ does not depend on location, the field is called *homogeneous*. The field of specific density of the body force, which possesses both of these properties is called *homogeneous gravitational field*.

Observations show that it is often permissible to regard the field of the body force density $\vec{b}$ as irrotational, *i.e.*, $\nabla \times \vec{b} = 0$. This greatly facilitates consideration and description of the body force. From the vector analysis it is known that for any twice differentiable scalar function φ the equality $\nabla \times \nabla \varphi = 0$ holds. Hence, a vector field $\vec{a}$, whose curl is identical zero, may be written as a gradient of some scalar quantity φ, which is called a *potential* of this vector field. In case of the body force density $\vec{b}$ one has

$$\vec{b} = -\nabla \Phi. \tag{7.12}$$

Here, the function Φ is the potential of the specific body force density $\vec{b}$. Such forces, in turn, are called potential forces[1]. The minus sign in (7.12) is a tribute to tradition. With this choice of the sign the gradient of Φ and the specific force density have $\vec{b}$ opposite directions. For example, the potential of gravity increases with height and the body force density vector is directed towards the center of the Earth. Henceforth, we shall always consider body forces as potential.

Finally, substituting (7.6), as well as (7.10) and (7.11) in (7.9), we obtain the *basic law of motion*:

$$d_t \mathbf{m} = \int_\chi \rho d_t \vec{v} dV = \int_{\partial \chi} \vec{\tau} dS + \int_\chi \rho \vec{b} dV. \tag{7.13}$$

This expression is also called the *momentum balance equation* (cf. with the general form of the balance equation (6.15 in Sec.6). The rate of change of momentum $d_t \mathbf{m}$ is balanced by the generation of momentum within the configuration χ *via* the body forces (second term) and the influx of momentum due to the momentum flux across the boundary $\partial \chi$ (first term).

The first step of deriving differential relation from the basic principle (7.6), thus, is completed: the force densities are defined. Now we need to rewrite the first term on the right-hand side in the form of an integral over the configuration χ.

7.3. THE EQUATION OF MOTION

In fluid mechanics the body forces are only of a secondary interest. The focus is on contact forces. Consider the stress field $\vec{\tau}$ on $\partial \chi$. What are those factors, vector $\vec{\tau}$ may depend on? Firstly, it should depend on the position of the point $\mathbf{x}$ and time t, *i.e.* on where it is located and when. Secondly, since $\vec{\tau}$ is a vector and $\partial\chi$ is a surface it is necessary to take into account their relative orientation[2].

A surface is called orientable if we may specify its inner and outer sides. The orientation of a surface at a point is usually associated with direction of a normal vector $\vec{n}$. By agreement, we build normal vector on the outer side of the surface and, thus, it marks the side which we call outer.

There exist non-orientable surfaces (the Möbius strip and the Klein bottle are classical examples of one-sided surfaces. Any vector, normal to such surface, may be superposed with a counter vector by moving one of them along the surface and normal to it (try to do this mentally, say, with the Möbius strip (see Fig. **7.1**)). However, this is exotics.

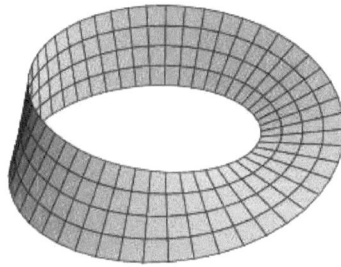

Fig. (7.1). The Möbius strip is an example of non-orientable surfaces.

We assume that the surface $\partial\chi$ is orientable, *i.e.* unambiguously associated with vector $\vec{n}$ normal to $\partial\chi$ at the point **x** (Fig. **7.2**), and directed outward with respect to $\chi(t, \mathcal{B})$. Now it is possible to take into account the effect of surface orientation on the stress acting on it. Repeating Cauchy's arguments, we accept the following

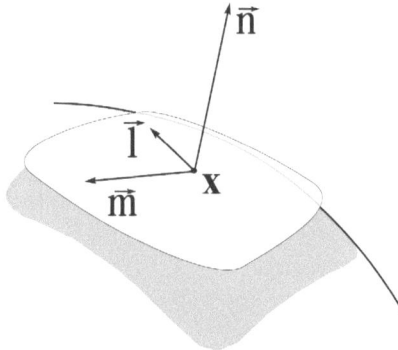

Fig. (7.2). Tangent vectors and a normal vector.

Postulate (Cauchy). Stress $\vec{\tau}$ at any given time t depends only on location $\mathbf{x}$ on the surface and orientation of the surface at this point

$$\vec{\tau} = \vec{\tau}(t, \mathbf{x}, \vec{n}). \tag{7.14}$$

The quantity $\vec{\tau}(t, \mathbf{x}, \vec{n})$ we interpret as a stress caused by external bodies at point $\mathbf{x}$ at time t and the quantity $\vec{\tau}(t, \mathbf{x}, -\vec{n})$, on the contrary, as a stress caused by the body $\mathcal{B}$ itself on a subset $\mathcal{B}_s^e$ (see, p.14). In accordance with the property 1 (the equality of absolute values of forces of action and reaction) of the system of forces, introduced on p.14, we have the following

Lemma (Cauchy):

$$\vec{\tau}(t, \mathbf{x}, \vec{n}) = -\vec{\tau}(t, \mathbf{x}, -\vec{n}). \tag{7.15}$$

Now it may be shown that the following theorem holds.

Theorem (Cauchy)[3]. The dependence of the stress on the normal $\vec{n}$ is linear, *i.e.* if $\vec{\tau}(t, \mathbf{x}, \vec{n})$ is a continuous function of time and place, then there exists a tensor $\mathrm{T}(t, \mathbf{x})$ such that

$$\vec{\tau} = \mathrm{T}\vec{n}. \tag{7.16}$$

The assertion that one vector is linearly dependent on another, in the component form means that each component of the first vector is linearly dependent on each component of the second vector. Let $\vec{\tau} = (\tau_1, \tau_2, \tau_3)$ and $\vec{n} = (n_1, n_2, n_3)$ with respect to some basis, then $\vec{\tau}$ will be linearly related to $\vec{n}$ if

$$\tau_i = \mathrm{T}_{ij} n_j, \quad \forall i = 1,2,3. \tag{7.17}$$

The coefficients T_{ij} are naturally regarded as components of a tensor T with respect to the same basis. In our case, the tensor T is called the *stress tensor* .

If at some point $\mathbf{x} \in \partial\chi$ we choose an orthonormal basis formed by the normal vector $\vec{n}$ and two arbitrary orthogonal vectors $\vec{l}$ and $\vec{m}$ tangent to $\partial\chi$ (see. Fig. **7.2**), then the stress $\vec{\tau}$ may be written as

$$\vec{\tau}(t, \mathbf{x}, \vec{n}) = \tau^l \vec{l} + \tau^m \vec{m} + \tau^n \vec{n}. \tag{7.18}$$

The coefficients τ^l, τ^m and τ^n *i.e.* the components of the vector $\vec{\tau}$ with respect to the basis $\{\vec{l}, \vec{m}, \vec{n}\}$, are projections of the stress on corresponding directions of the basis vectors and may be found using formulas:

$$\tau^l(t, \mathbf{x}) = (\vec{\tau}, \vec{l}), \quad \tau^m(t, \mathbf{x}) = (\vec{\tau}, \vec{m}), \quad \tau^n(t, \mathbf{x}) = (\vec{\tau}, \vec{n}).$$

The first two components τ^l and τ^m are projections of $\vec{\tau}$ on tangent basis vectors $\vec{l}$ and $\vec{m}$ and are called the *shear stresses*. The last component τ^n is the projection on the normal vector and is called the *normal stress* at point $\mathbf{x}$.

Looking at the expression (7.17) one may ask: what for all this? After all, to define the vector $\vec{\tau}$ three numbers are required, and to define the tensor T we need already nine. Why such a complication?

Here is why. Recall that our current goal is to derive a differential equation from the integral principle (7.13). However, we cannot follow the procedure described in Sec. 2.4.2 on p. 31, and merge integrals into one due to mismatched regions of integration. The stress tensor can help us to solve this problem. We replace the stress vector $\vec{\tau}$ in (7.13) with the stress tensor using the formula (7.16)

$$\int_\chi \rho d_t \vec{v} dV = \int_{\partial \chi} \mathrm{T} \vec{n} dS + \int_\chi \rho \vec{b} dV. \tag{7.19}$$

Now, after introduction of the stress tensor, it is possible to apply the Gauss' theorem to the problematic first term on the right-hand side (7.19) and rewrite it in the form

$$\int_{\partial \chi} \mathrm{T} \vec{n} dS = \int_\chi div \mathrm{T} dV, \tag{7.20}$$

and follow the described procedure. The equation (7.19) is written as follows

$$0 = \int_\chi \left(\rho d_t \vec{v} - div\mathrm{T} - \rho \vec{b}\right) dV. \tag{7.21}$$

We assume that the integrand is continuous and since the configuration is arbitrary, the equation holds, if and only if the integrand is equal to zero. Thus, we obtain the differential equation of balance of momentum density:

$$\rho d_t \vec{v} = div\mathrm{T} + \rho \vec{b}. \tag{7.22}$$

This equation is called the *equation of motion* of continuum[4].

In order to be able to use the law (7.22), it is necessary to define somehow the quantities in the right-hand side. If the body forces are generated only by the gravitation of the Earth, the vector $\vec{b}$ is the acceleration of gravity, and the problem of its description is solved by choosing a suitable basis (if one of the basis vectors coincides with $\vec{b}$ this vector has only one non-zero component) and measurement (of this single component).

With the stress tensor it is more complicated: it has nine components. It is possible to reduce the number of independent components by introducing additional relations between them. They are also taken from observations and measurements. One of the most general relations of this kind is the symmetry of the stress tensor. This assumption allows one to reduce the number of independent components to 6 and is made in the form of the

Postulate (Boltzmann[5]). *Stress tensor is symmetric*[6]

$$\mathrm{T} = \mathrm{T}^\mathrm{T}. \tag{7.23}$$

7.4. THE EQUATION OF MOTION OF THE PERFECT FLUID

We approached, in essence, the most important point in building a model of a moving fluid. A particular form of the stress tensor defines this model. Possible fluid models are even classified by the form of the stress tensor.

The Boltzmann postulate (7.23) has reduced the number of independent components of the stress tensor by three. Now they are six, but still not defined unambiguously. In order to select a particular form of the tensor T, and in general to understand what to choose from, we apply the trick, usual in such cases: we begin with the most simple form of the tensor and compare the result with nature, *i.e.*, we compare a solution of differential equations of the model with observational data.

If we assume that all calculations and measurements are performed correctly, the good agreement between calculations and observations means that the model in this case works well and the form of the stress tensor is quite suitable. The hope exists that this stress tensor will suit in other cases too. Otherwise the invented form of the stress tensor is no good and it is necessary to come up with another.

Apparently, the most simple meaningful hypothesis concerning the stress tensor may be considered its proportionality to the unit tensor[7]:

$$\mathrm{T} = -p\mathrm{I}. \tag{7.24}$$

In a model with such tensor T the stress vector $\vec{\tau}$ always has the same direction as the vector normal to the surface of the body in the study, and the shear stresses are absent. Indeed, substituting (7.24) in (7.16), we find

$$\vec{\tau} = \mathrm{T}\vec{n} = -p\mathrm{I}\vec{n} = -p\vec{n}. \tag{7.25}$$

Projection of $\vec{\tau}$ on the tangent plane is equal to zero and therefore the shear stresses are also zero. Thus, in the simplest case (7.24), we obtain a model of continuum which is able to describe only normal stresses.

Despite the specified feature, the selection is successful in the sense that, as observations show, there are many flows in which it is permissible to neglect shear stresses, and where the fluid model with the stress tensor of the form (7.24) gives results that are close to measurements. The proportionality factor p is called the *pressure*. The sign in (7.24) is selected so that for the positive pressure $p > 0$ the stress $\vec{\tau}$ acting on a closed surface, tends to compress the fluid contained inside. The pressure, specified at each point of the configuration χ, forms the pressure field. The body whose motion is determined by the dependence (7.24) is called perfect (or ideal) fluid.

Further, the divergence of T may easily be calculated and is equal to the pressure gradient:

$$(div(p\mathrm{I}))_i = \partial_{x_k}(p\delta_{ki}) = \partial_{x_i}p = \nabla_i p \quad \Rightarrow \quad div(p\mathrm{I}) = \nabla p. \quad (7.26)$$

Here the Kronecker symbol δ_{ik} is a component of the unit tensor. Substituting (7.24) into the expression (7.22), and taking into account (7.26), we obtain the equation of motion of the perfect fluid

$$\rho d_t \vec{v} = -\nabla p + \rho \vec{b}, \quad (7.27)$$

or the *Euler equation*. Another form of the same equation may be derived, if we write the derivative $d_t\vec{v}$ using the Euler formula (3.10):

$$d_t \vec{v} = \partial_t \vec{v} + (\vec{v}, \nabla)\vec{v}.$$

Substituting this expression in (7.27) and normalizing by the mass density, we get

$$\partial_t \vec{v} + (\vec{v}, \nabla)\vec{v} = -\frac{1}{\rho}\nabla p + \vec{b}. \quad (7.28)$$

7.5. THE EULER EQUATION IN COMPONENT FORM

Both forms (7.27) and (7.28) of the equation of motion of the perfect fluid, are vector equations. They are useful in constructing a theory, but when it comes to computing, the equations in component notation are necessary[8]. To obtain this form, we choose a coordinate system in the space of places. Among all possible coordinate systems it is better to choose the one that somehow takes into account certain peculiarities of the problem to be solved.

So, if a body force is due to gravity of the Earth and the vector $\vec{b}$ is always directed towards the center of the Earth, *i.e.*, down (since the direction "down" is just the direction of the vector $\vec{b}$), it is quite natural to use this specific direction for the selection and orientation of the coordinate system. Usually one of the coordinate axes is directed parallel to the vector $\vec{b}$. Then, if in each point **x** choose a coordinate basis $\{\vec{e}_i\}$, two components of the vector $\vec{b}$ will be zeros, and the third will be equal to the length of the vector, *i.e.* the gravity acceleration $|\vec{b}| = g$. The sign of this third component depends on whether co- or counter-directional are the base vector and the vector $\vec{b}$. For example, if we select spherical geocentric coordinates (λ, φ, r) with the axis r directed against the vector $\vec{b}$ then $\vec{b} = (0, 0, -g)$. On the contrary, if we consider the problem in Cartesian coordinates (x, y, z) with the axis z directed vertically downward, then $\vec{b} = (0, 0, g)$.

So, let the coordinate system in the space of places is chosen, and at each point of configuration of the body the coordinate basis $\{\vec{e}_i\}$ is defined, with respect to which

$$\vec{v} = v_i \vec{e}_i, \quad \vec{b} = (0,0,-g) = -g\delta_{3i}\vec{e}_i, \quad \nabla(\cdot) = \vec{e}_i \nabla_i(\cdot). \tag{7.29}$$

The Kronecker symbol in the expansion of the vector $\vec{b}$ allows one to take it into account only in the third equation. Substituting (7.29) in (7.28), we obtain

$$\partial_t(v_i \vec{e}_i) + (\vec{v}, \nabla)(v_i \vec{e}_i) = -\frac{1}{\rho}(\nabla_i p)\vec{e}_i - g\delta_{3i}\vec{e}_i, \tag{7.30}$$

or

$$\left(\partial_t v_i + (\vec{v}, \nabla)v_i + \frac{1}{\rho}\nabla_i p + g\delta_{3i}\right)\vec{e}_i = 0. \tag{7.31}$$

The right side is the zero vector, which components are zeros with respect to arbitrary basis. This means, that equality (7.31) holds if and only if terms in parentheses are equal to zero for each i. Thus, we get three scalar equations for velocity components which correspond to one vector equation (7.28)

$$\partial_t v_i + (\vec{v}, \nabla)v_i = -\frac{1}{\rho}\nabla_i p - g\delta_{3i}. \tag{7.32}$$

Further we may calculate the scalar product $(\vec{v}, \nabla) = v_k \nabla_k$ and finally the Euler equations (now they are three) take the form

$$\partial_t v_i + v_k \nabla_k v_i = -\frac{1}{\rho}\nabla_i p - g\delta_{3i}. \tag{7.33}$$

Let us keep in mind that

1. k is a dummy index which indicates the summation from 1 to 3;
2. i is a free index; it takes the values 1, 2 and 3, different in different equations, and therefore it numbers these equations;
3. ∇_k is a covariant derivative. When the selected coordinate system is Cartesian, the covariant derivative is equal to partial derivative $\nabla_k v_i = \partial_{x_k} v_i$ and the covariant derivative of a scalar is always equal to partial derivative $\nabla_i p = \partial_{x_i} p$. In this case, the equation (7.33) takes the form

$$\partial_t v_i + v_k \, \partial_{x_k} v_i = -\frac{1}{\rho} \partial_{x_i} p - g \delta_{3i}.$$

If the coordinate system is non-Cartesian (*e.g.*, spherical), then some of the Christoffel symbols are non-zero, and should be taken into account in the equations.

7.6. SUMMARY

Our fluid model now contains two equations: the continuity equation and the equation of motion. However, in general case, the system of equations is still not closed. A new unknown scalar function p pressure has appeared in the equation of motion.

The system of equations written in component form contains five unknowns (ρ, $\vec{v}$ and p). If one of the fields (ρ or p), is specified the model equations allow calculating the velocity field and the second scalar field. What is to be done in general case? Apparently, try to find another independent equation relating the variables of the model and, thus, close the system. We shall try to do this in the next chapter.

NOTES

[1] Introduction of a potential and representation of a vector quantity in the form (7.12) is useful not only regarding forces, but also in other cases (see, *e.g.*, Sec.10.3).

[2] Probably everyone has experienced himself the influence of orientation of a surface. It is easy to stand on a surface, which is perpendicular to gravity, and to remain on one's feet on rather steep slope of a roof or a hill is already a problem.

[3] For the proof, see any textbook listed in the bibliography.

[4] C.Truesdell calls this equation the Cauchy's *First Law of Motion* [1].

[5] Boltzmann Ludwig (1844–1906), an Austrian physicist.

[6] C.Truesdell calls this postulate [1] the Cauchy's *Second Law of Motion*, while J.Serrin [4] calls it the Boltzmann postulate.

[7] The zero tensor is obviously not suitable: it eliminates all of the contact forces completely.

[8] Here we are acting exactly as before on the p.39.

CHAPTER 8

Energy

Abstract: We postulate the total energy balance (the first law of thermodynamics) and derive the third equation of the fluid model, the internal energy density balance equation. The kinetic energy of the body is defined and its balance is discussed. The power of forces acting on the body is also discussed. The notion of internal energy is introduced, and its balance is considered. Finally, we formulate the complete system of equations of the fluid model.

Keywords: Energy balance, Equation of state, First law of thermodynamics, Force power, Heat flux, Internal energy, Kinetic energy, Mechanical energy, Potential energy, System of fluid model equations, Total energy.

8.1. KINETIC ENERGY AND ITS BALANCE

While studying integral parameters and corresponding densities we introduced the mass density, the scalar field defined on points of a body. This is not the only scalar field, which may be defined. In general case, points of a body move along their trajectories with velocity vector $\vec{v}$ and beside ρ an absolute value of $\vec{v}$ is also defined at each point. A quantity proportional to the square of the absolute value of velocity

$$k = \tfrac{1}{2}\rho|\vec{v}|^2 = \tfrac{1}{2}\rho(\vec{v}, \vec{v}) \qquad (8.1)$$

is called the *kinetic energy density* and corresponding integral parameter $K(t, \mathcal{B})$ is called the *kinetic energy* (*i.e.*, the energy of motion) of the body at time t

$$K(t, \mathcal{B}) = \int_{\chi} k \, dV. \qquad (8.2)$$

To understand what determines the evolution of the kinetic energy, let's find the rate of its change. The time derivative of K gives

$$d_t K \overset{(7.4)}{=} \int_\chi \rho d_t \frac{k}{\rho} dV \overset{(8.1)}{=} \frac{1}{2} \int_\chi \rho d_t (\vec{v}, \vec{v}) dV = \int_\chi \rho (\vec{v}, d_t \vec{v}) dV. \qquad \textbf{(8.3)}$$

Here in the first equality we have used the auxiliary formula (7.4) derived on p.83, and in the third one we have taken into account that $d_t(\vec{v}, \vec{v}) = 2(\vec{v}, d_t\vec{v})$. As a result, under the usual assumptions, we obtain the differential equation which corresponds to (8.3):

$$\rho d_t \frac{k}{\rho} = \rho(\vec{v}, d_t \vec{v}) = (\vec{v}, \rho d_t \vec{v}). \qquad \textbf{(8.4)}$$

Another way to derive this relation consists in differentiating of definition (8.1). However, in this case the interconnection with $d_t K$ would have been somewhat obscure.

So, to find the rate of change of the specific kinetic energy density $d_t \frac{k}{\rho}$ and, then, the rate of change of the kinetic energy of the body $d_t K$ we need to calculate the scalar product of the velocity vector $\vec{v}$ and the acceleration $d_t \vec{v}$. The acceleration may be found from the equation of motion (7.22), left-hand side of which is equal to $\rho d_t \vec{v}$. Thus, substituting the equation of motion into (8.4), we find

$$\rho d_t \frac{k}{\rho} = (\vec{v}, div\mathrm{T} + \rho \vec{b}) = (\vec{v}, div\mathrm{T}) + (\vec{v}, \rho \vec{b}). \qquad \textbf{(8.5)}$$

This expression is called the *equation of the specific kinetic energy density balance*[1]. In particular, for the perfect fluid (T= –*p*I) we get

$$\rho d_t \frac{k}{\rho} = -(\vec{v}, \nabla p) + \left(\vec{v}, \rho \vec{b}\right)$$

or in the component form (and normalizing by ρ)

$$\partial_t \frac{k}{\rho} + v_i \partial_{x_i} \frac{k}{\rho} = -\frac{1}{\rho} v_i \partial_{x_i} p + v_i b_i.$$

The rate of change of the kinetic energy of the body in the general case, may be found by integrating (8.5) over points of the current configuration χ:

$$d_t K = \int_\chi (\vec{v}, div\mathrm{T} + \rho\vec{b})dV = \int_\chi (\vec{v}, div\mathrm{T})dV + \int_\chi (\vec{v}, \rho\vec{b})dV. \qquad \textbf{(8.6)}$$

What determines the change in kinetic energy, *i.e.*, what stands on the right side of (8.6)? Once the left side is the rate of change of energy, and energy is the body's ability of doing work, then, on the right side, by definition, should be the rate of doing work by the system of forces, *i.e.* the power of forces. Let's see what forces and how affect the change of the kinetic energy.

In the beginning, we define the power of the system of forces. Let $\vec{v}$ be the velocity of points of the body under the action of a force **f** with the density $\vec{f}$, then the integral

$$W = \int_A (\vec{v}, \vec{f})dA$$

will be called the *power of the force* **f**. If **f** is a body force, the integration is performed over points of configuration χ of the body, and if **f** is a contact force, then the integral is calculated over points of the border of configuration $\partial\chi$. Now we calculate the power of the system of forces acting in our case

$$W = \underbrace{\int_\chi (\vec{v}, \rho\vec{b})dV}_{W_B} + \underbrace{\int_{\partial\chi} (\vec{v}, \vec{\tau})dS}_{W_C}. \qquad \textbf{(8.7)}$$

The first term W_B is the rate of doing work by body forces, and the second W_C is the rate of doing work by contact forces.

To compare this result with the expression (8.6), we transform W_C to the integral over the volume of configuration. Firstly, we write down stresses in terms of the stress tensor

$$W_C = \int_{\partial\chi} (\vec{v}, \vec{\tau})dS = \int_{\partial\chi} (\vec{v}, T\vec{n})dS = \int_{\partial\chi} (T\vec{v}, \vec{n})dS. \qquad \textbf{(8.8)}$$

Here, in the latter equality we have used the formula (4.2) and the Boltzmann postulate (7.23). Further, applying the Gauss' theorem to (8.8), we find the power of contact forces:

$$W_C = \int_{\partial\chi} (T\vec{v}, \vec{n})dS = \int_{\chi} div(T\vec{v})dV. \qquad \textbf{(8.9)}$$

The total power of forces acting on the body is equal to

$$W = \int_{\chi} \left(div(T\vec{v}) + (\vec{v}, \rho\vec{b})\right)dV. \qquad \textbf{(8.10)}$$

Comparing this expression with the right-hand side of (8.6), we see that 1) the work of the body forces consists solely in the change of the kinetic energy of a body and 2) the contact forces also change the kinetic energy, but in addition they do something else. What?

Let's calculate the divergence $div(T\vec{v})$ using as an intermediate expression in the chain of reasoning, the component notation:

$$div(T\vec{v}) = \nabla_k\left(T_{kj}v_j\right) = \left(\nabla_k T_{kj}\right)v_j + T_{kj}\nabla_k v_j = (\vec{v}, divT) + T:\nabla\vec{v}. \qquad \textbf{(8.11)}$$

The first term in the last expression of (8.11) is the part of the power of contact forces, which changes the kinetic energy. What does another part do?

Consider the inner product $T: \nabla\vec{v}$. In case of the perfect fluid, when $T = -pI$, we have

$$T: \nabla \vec{v} = -p \delta_{ij} \nabla_j v_i = -p \nabla_i v_i = -p(\nabla, \vec{v}), \tag{8.12}$$

The quantity $(\nabla, \vec{v})$ is determined by the relative compressibility of continuum (see Eq. (6.19)) and, consequently,

$$T: \nabla \vec{v} = \frac{p}{\rho} d_t \rho. \tag{8.13}$$

Now it is clear, that the second part of the power of contact forces is spent on the change of volume due to uniform compression/expansion (after all, only normal stresses are non-zero, while shear stresses by virtue of (7.25) are absent), and therefore does not affect the change of kinetic energy of the body.

Thus, the situation is possible when contact forces act, but do not participate in the change of kinetic energy of the body. To see this more clearly, we shall get rid of the body forces on the right-hand side, introducing the concept of *mechanical energy*. We assume (as we have previously agreed) that the body force is potential with time-independent potential Φ. Then the term $(\vec{v}, \rho \vec{b})$ in the equation (8.5) may be transformed as follows

$$(\vec{v}, \rho \vec{b}) = -\rho(\vec{v}, \nabla \Phi) - \rho \underbrace{\partial_t \Phi}_{=0} = -\rho d_t \Phi, \tag{8.14}$$

and the equation (8.5) itself may be written in the form

$$\rho d_t \left(\frac{k}{\rho} + \Phi \right) = (\vec{v}, div T). \tag{8.15}$$

The expression in parentheses in the left-hand side is called the *specific density of mechanical energy*, and the equation (8.15) is called the *balance equation of mechanical energy density*.

Exercise. Derive the equation (8.15).

Integrating (8.15) over χ we find

$$d_t(K + U) = \int_\chi (\vec{v}, div\mathrm{T})dV. \tag{8.16}$$

The quantity $U = \int_\chi \rho \Phi dV$ is called the *potential energy* of configuration χ of the body $\mathcal{B}$ and the sum $(K + U)$, in turn, is called the *mechanical energy* of the body.

Thus, if $(\vec{v}, div\mathrm{T}) = 0$, contact forces are not involved in the change of mechanical energy, accomplishing only a uniform compression. In this case the mechanical energy of the body is conserved during the motion. The potential and kinetic energies are not necessarily conserved, and they may exchange with each other with portions of energy. However, if K (or U) does not change, the other one does not change too. If the fluid is perfect the equality $(\vec{v}, div\mathrm{T}) = 0$ means that $\nabla p \perp \vec{v}$ *i.e.* the velocity vector is orthogonal to the pressure gradient. Such situation occurs, for example, in a plane-parallel flow of a horizontal layer of fluid in a gravitational field. The pressure gradient is directed vertically downward perpendicular to the flow velocity and does not affect its value. In other words, it does not cause variations of the kinetic energy.

8.2. THE INTERNAL ENERGY AND ITS BALANCE

Consider again the total power of forces acting on the body (8.10)

$$
\begin{aligned}
W &= \int_\chi \left(div(\mathrm{T}\vec{v}) + (\vec{v}, \rho\vec{b}) \right)dV = \\
&= \int_\chi (\vec{v}, div\mathrm{T} + \rho\vec{b})dV + \int_\chi (\mathrm{T}:\nabla\vec{v})dV.
\end{aligned}
\tag{8.17}
$$

The first term on the right side of (8.17) is the rate of change of kinetic energy $d_t K$. Obviously, the last term has the same nature as the first one. Both have one and the same source, a power of forces acting on the body. By analogy with the first term we shall interpret the second one as the rate of change of some other kind of energy. We denote it by the letter E and call the *internal energy*. The internal energy density will be denoted by ε, *i.e.*, we define

$$E = \int_\chi \varepsilon dV. \tag{8.18}$$

With this interpretation of the latter term in (8.17), all power of forces is spent on the change of energy and the rate of change of the internal energy of continuum is equal to:

$$d_t E = \int_\chi (\mathrm{T} : \nabla \vec{v}) dV. \tag{8.19}$$

The energy balance now looks more complete:

$$d_t(K + E) = W. \tag{8.20}$$

The change of the overall energy $(K+E)$ is determined by the power of forces, and is equal to zero when forces are absent. The quantity $(K+E)$ is called the *total energy* of the body, and $(k+\varepsilon)$ is called the *total energy density*, respectively. Along with the mass, the total energy is another one conserved scalar quantity.

What does the internal energy E actually describe? Or is it all just a trick to make a well-shaped theory, and which has no physical meaning? Fortunately physical justification exists, and this new setting allows us to introduce in the model a completely unexpected characteristic of the moving continuum. It is connected with that we feel as heat, the heating of continuum.

It is known that if, say, a gas is compressed, it warms up, and *vice versa* , while expanding it cools down. In both cases, the kinetic energy of fluid particles changes. However, these changes occur on a microscopic scale, at the level of molecules, where the individual description becomes impossible. Refusing this way, and accepting the continuity hypothesis, we use the average, parametric description. The internal energy and its density are those parameters that characterize the average kinetic energy of molecules. The first parameter is an integral characteristic of the body, and the second is a local characteristic of a point of continuum.

The last thing that is left to do in the energy balance (8.20) is to consider the obvious fact that the heat may not only be produced by changing a configuration of a body, but it is also possible to interchange it with the environment. The final assertion is formulated in the form of a postulate known as the *first law of thermodynamics*

$$d_t(K + E) = W + Q, \tag{8.21}$$

which is just the *total energy balance equation*. The meaning of notations is the same as in (8.20), and Q is the rate of heating of the body. The quantity $(W+Q)$ in the right side is the total energy source power. If the source is absent the rate of change of the total energy during the motion is zero, *i.e.* total energy is conserved.

Now using this integral relation let's try to derive a differential equation. By analogy with forces, consider two variants of the heat transfer: a *body heating* Q_B due to radiation and a *contact heating* Q_C through conduction. The total heating rate is considered to be the sum of Q_B and Q_C.

$$Q = Q_B + Q_C. \tag{8.22}$$

The quantity Q_B relies to be a continuous function of the volume of configuration of the body:

$$Q_B = \int_\chi \rho s\, dV, \tag{8.23}$$

(the density ρs is a *heat flux due to radiation*), and the quantity Q_C relies to be a continuous function of the surface area $\partial \chi$ bounding the body:

$$Q_C = \int_{\partial \chi} q\, dS. \tag{8.24}$$

Here q is a heat influx by means of conduction[2]. Observations show that the analogy with forces goes further and allows us to formulate assertions similar to the Cauchy's postulate and theorem. Namely, assume that are valid the following

Postulate

$$q = q(t, \mathbf{x}, \vec{n}), \tag{8.25}$$

which states that the heat flow q at any given time t depends only on the position $\mathbf{x}$ of a point and on the surface orientation $\vec{n}$ at this point, and

Theorem (the *Fourier-Stokes[3] principle of heat flow*)

$$q(t, \mathbf{x}, \vec{n}) = (\vec{h}(t, \mathbf{x}), \vec{n}), \tag{8.26}$$

asserting that the dependence of the heat flow on the normal vector is linear. The quantity $\vec{h}(t, \mathbf{x})$ is called the *heat flux density*. Combining (8.22)–(8.26), we obtain

$$Q = \int_\chi \rho s \, dV + \int_{\partial\chi} (\vec{h}, \vec{n}) \, dS.$$

Applying further, the Gauss' theorem to the second term, we find

$$\int_{\partial\chi} (\vec{h}, \vec{n}) \, dS = \int_\chi (\nabla, \vec{h}) \, dV \quad \Rightarrow$$
$$\Rightarrow \quad Q = \int_\chi \left(\rho s + (\nabla, \vec{h}) \right) dV. \tag{8.27}$$

To derive the differential balance equation we have to put the pieces together, *i.e.*, substitute expressions (8.16), (8.18) to the left-hand side, and expressions (8.25), (8.27) to the right-hand side of the first law of thermodynamics (8.21):

$$d_t \int_\chi (k + \varepsilon) \, dV = \underbrace{\int_\chi \left(div(\mathrm{T}\vec{v}) + (\vec{v}, \rho\vec{b}) \right) dV}_{W} + \underbrace{\int_\chi \left((\nabla, \vec{h}) + \rho s \right) dV}_{Q}. \tag{8.28}$$

Calculating the derivative on the left-hand side using the formula (7.4), combining integrals and assuming the continuity of integrand and arbitrariness of configuration, we obtain the balance equation of the total energy density:

$$\rho d_t \frac{k+\varepsilon}{\rho} = div(\mathrm{T}\vec{v}) + (\vec{v}, \rho\vec{b}) + (\nabla, \vec{h}) + \rho s. \tag{8.29}$$

Now we can get the *internal energy balance*. This easily may be done subtracting, for example, (8.6) or (8.5) from (8.28) or (8.29), respectively. We shall subtract (8.6) from (8.28). In view of formula (8.11), we have:

$$d_t E = \int_\chi \rho d_t \frac{\varepsilon}{\rho} \, dV = \int_\chi (\mathrm{T} : \nabla\vec{v}) \, dV + \int_\chi \left((\nabla, \vec{h}) + \rho s \right) dV. \tag{8.30}$$

This result is consistent with the expression (8.19), which does not take into account the external heating. The differential equation for ε may easily be obtained directly from (8.30). Here it is

$$\rho d_t \frac{\varepsilon}{\rho} = \mathrm{T}:\nabla\vec{v} + (\nabla,\vec{h}) + \rho s. \tag{8.31}$$

This equation describes the rate of change of the specific density of the internal energy of the body under the contact forces $(\mathrm{T}:\nabla\vec{v})$ acting on its surface, and the external heating $(\nabla,\vec{h})+\rho s$. In case of the perfect fluid (see the formula (8.13)) and in the absence of external heating, the equation (8.31) takes the form:

$$d_t \frac{\varepsilon}{\rho} - \frac{p}{\rho^2} d_t\rho = d_t\frac{\varepsilon}{\rho} + p d_t\frac{1}{\rho} = 0. \tag{8.32}$$

This expression for the first law of thermodynamics may often be found in the literature. Finally, if we consider fluid incompressible (*i.e.*, assuming $d_t\rho{=}0$), the internal energy is conserved at each point of the body (fluid): $d_t\varepsilon = 0$.

8.3. SUMMARY

Recall how and why we began this chapter. We have had a non-closed system of two equations. It was necessary to derive an equation that would close it, *i.c.*, an equation that describes some new connection between existing variables. What have we gained?

We got as many as four equations (balance of kinetic, mechanical, total, and internal energies). However, the new is only one: it is the balance equation either of the total energy, or the internal energy (any of them is suitable). The other three are not independent and may be easily obtained from the selected one and already existing equations.

Our fluid model now includes five independent scalar equations, but is still non-closed. The new equation, which was needed to close the former system, again adds an unknown quantity, the internal energy density. A vicious circle: each new equation adds a new unknown variable and the system remains non-closed.

Nothing is left but to find empirically a relation between the existing unknowns of the problem. However, a universal experiment cannot be conducted: any experiment requires a particular fluid. Therefore, if the above-mentioned relation would be found, it is likely to be applicable only to the fluid, for which it was found. Thus, here our theory stops to be universal.

8.4. THE EQUATION OF STATE

At this step of developing of the model of continuum we have following unknowns ρ, $\vec{v}$, p, ε. In order to simplify the problem of searching missing relation between these functions we accept additional hypotheses.

1. Firstly, it is a hypothesis of *local thermodynamic equilibrium*, which means that all equations of equilibrium thermodynamics are valid for infinitesimal elements of mass of nonequilibrium systems. Currently it is assumed [6] that this hypothesis is always applicable except in special cases of fast processes. Thermodynamic equilibrium (like any other equilibrium) means an arbitrarily long preservation of values of characteristics of a body under constant external conditions. In this case, the time derivatives of thermodynamic characteristics locally are equal to zero.
2. Secondly, here we consider only the simplest case of a single-component and a single-phase medium. In other words, a body under consideration consists of one substance in one thermodynamic state.
3. Thirdly, we assume that our environment as a thermodynamic system does not depend on values of a macroscopic velocity and is a two-parameter system. This means that its state is completely determined by any two parameters.

Thus, taking into account all three assumptions, it turns out that the remaining unknowns ρ, p and ε should be bound by one relation. This one, for example,

$$\rho = \rho(p, \varepsilon). \tag{8.33}$$

Expressions like (8.33) are called *equations of state*. They are algebraic, not differential and relate the values of the variables themselves, and not their derivatives.

The mass density in the equation (8.33) is a composite function of time and place:

$$\rho(t, \mathbf{x}) = \rho\big(p(t, \mathbf{x}), \varepsilon(t, \mathbf{x})\big). \tag{8.34}$$

Accordingly, any derivative of the mass density may be expressed in terms of derivatives of p and ε

$$\partial_{x_\alpha}\rho = \partial_p\rho\,\partial_{x_\alpha}p + \partial_\varepsilon\rho\,\partial_{x_\alpha}\varepsilon. \tag{8.35}$$

This equality is often used in transformations of the model equations.

Moving in the indicated direction, it is possible to achieve some success. By processing numerous experiments a number of equations of state (often quite complex) were obtained for various substances. Each of them is the last relation, closing the system of the fluid mechanics equations. They add missing connection without increasing the number of unknowns. This is achieved, however, at the cost of lost of universality of the model. Until now, all the equations were valid for any continuum (under the assumptions made). Adding a particular equation of state makes the system valid only for the substance to which this equation corresponds.

8.5. MODEL OF THE FLUID: DISCUSSION OF THE RESULTS

Thus, out of the four laws of nature (or axioms[4])

- the mass conservation law (6.16),
- the basic principle of dynamics (7.3),
- the total energy conservation law or the first law of thermodynamics (8.21) and
- the equation of state[5] (8.33),

using some additional hypotheses and assumptions we have been able to obtain a closed system of differential equations, which are referred to as a *mathematical model of fluid dynamics*.

Before finishing this section, we are going to answer one possible question: where from the internal energy (heat) suddenly had appeared? Nothing like that has been introduced in the beginning. Not exactly so.

In discussing the continuity hypothesis, it was noted that the adoption of this hypothesis means our intention to describe explicitly only macroscopic processes. If we were going to describe the motion of molecules, how we could ignore the discreteness of matter? Since we regarded it insignificant, this has allowed us to consider the matter as continuum.

To what extent is this not essential? May we neglect it at all, or still need to somehow take into account by introducing, for example, a special parameter? Observations show that, depending on the specific situation (fluid or scales of phenomena) one can do this way and that.

If errors associated with neglecting of molecular processes are too large, one may try to take them into account parametrically, *i.e.* introduce a function of time and space $\varepsilon(t, \mathbf{x})$, interpret it as a density of the average kinetic energy of molecular motions and call it an *internal energy*. This was done when writing the equation (8.33). The fluid model, wherein the equation of state is written in the form (8.33), is called *baroclinic*. Consideration of the internal energy here is a necessary part of the problem. It ensures the law of conservation of total energy, since in the general case, the kinetic energy is not conserved. Loss of kinetic energy is balanced by increasing of internal energy (heating of continuum) and thus the knowledge of this parameter is required for correct description of the fluid dynamics.

On the contrary, if observations show that the losses of kinetic energy due to internal friction forces are negligible, and the macroscopic fluid dynamics depends weakly on the processes occurring at the molecular level, then the two-parameter hypothesis (8.4) may be replaced by a stronger one-parameter assumption and one may write the equation of state in the form:

$$\rho = \rho(p), \tag{8.36}$$

i.e. to consider the mass density independent on internal energy. Such a model of the fluid is called *barotropic*. In this case the description of motion of the fluid does not require consideration of internal energy.

In extreme case it is possible to invent more complicated relationships, giving up any of hypotheses, accepted at the beginning of the paragraph, but usually the expressions like (8.33) are quite sufficient.

NOTES

[1]Note that here we have used the equation of motion in the form of the Cauchy's law of motion and, hence, the equality (8.5) holds for an arbitrary fluid, which motion is described by the law of motion (7.22), and not necessarily perfect fluid.

[2]As in case of contact forces (see, p.88), one should not think that the heat transfer by thermal conductivity takes place only on the surface of a body. This process occurs inside as well. However, at any interior point the heat influx is compensated by its outflow to the points of the same body and remains uncompensated in the points on the surface of the body only.

[3]Fourier Jean Baptiste Joseph (1768–1830), a French physicist and mathematician. Stokes George Gabriel (1819–1903), an English physicist and mathematician.

[4]Note, that inasmuch as we are developing a mathematical theory, it should be based on a sufficient number of axioms. From the mathematical point of view the totality of these axioms is not unique. However, since here we are interested in a theory which is capable of describing some physical phenomenon, *i.e.* actually is a model of this phenomenon, we must select these axioms such that they correspond to the so-called laws of nature, *i.e.* empirical dependencies of the most general type.

[5]This relationship, as the first three, has been extracted from experience and is also a statement accepted without proof, *i.e.* is the axiom or the law of nature.

Part II
Applications of the Fluid Model

CHAPTER 9

Perfect Fluid

Abstract: We study the model of the perfect fluid. Firstly, we discuss the problem posing for this fluid model and derive the Lamb form of the equation of motion. Further we investigate some simplest problems which are associated with the perfect fluid. We consider the hydrostatics, the barotropic fluid, the Bernoulli equation, trajectories and streamlines. Finally we obtain the vorticity equation.

Keywords: Barotropic fluid, Bernoulli equation, Hydrostatics, Lamb form of equation of motion, Perfect fluid, Stationary flow, Streamlines, Trajectories, Vortex equation.

9.1. INTRODUCTION

9.1.1. Problem Posing in the Perfect Fluid Mechanics

Now we have the closed system of differential equations describing the perfect fluid motion. Here it is:

$$\partial_t \rho + (\nabla, \rho \vec{v}) = 0, \tag{9.1}$$

$$\partial_t \vec{v} + (\vec{v}, \nabla)\vec{v} = -\frac{1}{\rho}\nabla p + \vec{b}, \tag{9.2}$$

$$d_t \frac{\varepsilon}{\rho} + p d_t \frac{1}{\rho} = \frac{1}{\rho}(\nabla, \vec{h}) + s, \tag{9.3}$$

$$\rho = \rho(p, \varepsilon). \tag{9.4}$$

Three differential equation bind four unknown functions ρ, $\vec{v}$, p and ε. At that, the system comprises one vector equation (the equation of motion (9.2)), and the totality of unknowns comprises vector ($\vec{v}$), respectively. The quantity $\vec{b}$ is considered a parameter of the problem. The sources of internal energy either are assumed to be known (s) or are expressed in terms of the sought quantities $(\vec{h})$, as it will be shown later. The latter, fourth equation allows exclusion of one of the scalar functions (*e.g.*, p) out of the number of unknowns.

The relations (9.1)–(9.3) make the system of three-dimensional evolutionary differential equations of the 1st order. To find a particular solution it is necessary to formulate a mixed initial-boundary value problem. Since all differential equations are evolutionary[1] one needs to specify, as the initial conditions, values of the fields ρ, $\vec{v}$ and ε at the initial time $t = 0$. At the boundary of the range of definition of the unknown functions the boundary conditions for ρ, $\vec{v}$, and ε should be specified.

Once the equations are written down, the range of definition of the functions is selected and the initial and boundary conditions are given, one may say that the *problem is posed*.

9.1.2. The Equation of Motion in the Lamb Form

When considering the dynamics of the perfect fluid it may be useful to write the Euler equation of motion (9.2) in the so-called Lamb form[2]. Its peculiarity consists in isolation of a term, which describes the vortex motion of fluid, in the left-hand side of the equation. To obtain this form, we write the second term in the left-hand side of (9.2) in terms of the velocity gradient $G = \nabla \vec{v}$ and the spin $W = \frac{1}{2}(G - G^T)$:

$$(\vec{v}, \nabla)\vec{v} = G\vec{v} = (G - G^T)\vec{v} + G^T\vec{v} = 2W\vec{v} + G^T\vec{v}.$$

The first term, in accordance with the formula (5.19), gives $2W\vec{v} = \vec{\omega} \times \vec{v}$ or $2W\vec{v} = -\vec{v} \times \vec{\omega}$. Here $\vec{\omega} = \nabla \times \vec{v}$ is the vorticity vector. The expression for j-th component of the vector $G^T\vec{v}$ easily may be found

$$(G^T\vec{v})_j = (\partial_{x_j}v_k)v_k = \frac{1}{2}\partial_{x_j}(v_k^2),$$

and therefore $G^T\vec{v} = \frac{1}{2}\nabla|\vec{v}|^2$. Thus,

$$(\vec{v}, \nabla)\vec{v} = \frac{1}{2}\nabla|\vec{v}|^2 - \vec{v} \times \vec{\omega}. \tag{9.5}$$

Substituting this expression into the equation (9.2), we obtain

$$\partial_t \vec{v} - \vec{v} \times \vec{\omega} + \frac{1}{2} \nabla |\vec{v}|^2 = -\frac{1}{\rho} \nabla p + \vec{b}.$$

Finally, regrouping terms and assuming, as usual, that the body force $\vec{b}$ is potential with the potential $-\Phi$, we write

$$\partial_t \vec{v} + \nabla \left(\frac{1}{2} |\vec{v}|^2 + \Phi \right) = -\frac{1}{\rho} \nabla p + \vec{v} \times \vec{\omega}. \tag{9.6}$$

This is the *equation of motion in the Lamb form*.

9.1.3. Remarks on Applications of the Model

Solving problems of the fluid mechanics in general is a very laborious and sometimes unnecessary process. Often, using a priori information about the sought solution, it is possible to simplify considerably the original equations, and in rare cases even obtain an analytical solution.

At the same time it is worth to keep in mind that after the model of the moving continuum was formulated, we do not plan derivation of other, fundamentally new equations. The base model of continuum is all the time the same: the mass conservation law (6.16), the basic principle of dynamics (7.3) and the first law of thermodynamics (8.21). The only significant change we shall afford to do is the redefinition of the stress tensor. This is necessary to resort in cases when the existing base model of the perfect fluid ceases to describe the observed flows, and needs to be improved. We will make use of this possibility when considering viscous and turbulent continuum.

Much more often it will be sufficient to neglect something in the already constructed model to simplify existing complex system of equations, adapting it for the description of specific flow regimes. Just this kind of simplifications we are going to use further. We sacrifice universality for the opportunity to describe easier any particular case. Please note that whenever we shall deduce new equations (system of equations of hydrostatics in the Sec. 1.2, the vorticity equation in the Sec. 1.3.3, *etc.*), we will not act the way we did in the derivation of equations of the model of continuum in the first part of the book, *i.e.*, we won't start with a discussion of the fundamentals, basic assumptions, but using the already developed model we shall simplify something in it (either we equate velocities or derivatives to zero, or the equation of state we write in abbreviated form and the like). The same way we shall proceed now when considering one of these simple particular cases, which is the hydrostatic equilibrium.

9.2. HYDROSTATICS

The hydrostatics describes the mechanical equilibrium of a *fluid at rest*. In this case, $\vec{v} = 0$ and equations (9.1 - 9.3) take the form:

$$\partial_t \rho = 0, \tag{9.7}$$

$$0 = -\nabla p + \rho \vec{b}, \tag{9.8}$$

$$\partial_t \frac{\varepsilon}{\rho} = -p\, \partial_t \frac{1}{\rho} + \frac{1}{\rho}(\nabla, \vec{h}) + s. \tag{9.9}$$

The continuity equation (9.7) implies conservation of the mass density at each point of a fluid at rest. Changes in internal energy, as show the equations (9.7) and (9.9), are only due to external heating. The equation (9.8) allows one to find the pressure distribution in this case.

Let's choose the Cartesian basis, direct the basis vector $\vec{e}_3$ straight upward, and denote $x_1 = x$, $x_2 = y$, $x_3 = z$. Then $\vec{b} = (0, 0, -g)$, and the equation (9.8) is equivalent to three scalar equations:

$$\begin{aligned} -\partial_x p &= 0, \\ -\partial_y p &= 0, \\ -\partial_z p &= \rho g. \end{aligned} \tag{9.10}$$

The first two equations imply that $p = p(z)$. For the third equation it is possible to formulate the Cauchy problem, if we specify a pressure value at some z.

Suppose, for example, that the origin is located at the lower boundary of the liquid layer of depth h and the pressure at the surface $z = h$ is equal to p_0. Then integration of the equation (9.10) gives

$$\int_{p(z)}^{p(h)} dp = -g \int_z^h \rho\, dz.$$

If it is possible to assume $\rho = const$ then taking into account the condition $p(h) = p_0$ we obtain:

$$p = p_0 + \rho g(h - z),$$

i.e. pressure increases linearly with depth and reaches the highest value $p = p_0 + \rho gh$ at the bottom, where $z = 0$.

9.3. BAROTROPIC MODEL

If dynamics of a fluid is described using barotropic model, the first term on the right side of the equation of motion (9.2) may be written as the gradient of some function. Such representation allows one to simplify the model of continuum and to obtain important information about the nature of the flow.

We define the function P *via* the relation:

$$P = \int \frac{1}{\rho} dp.$$

By virtue of the fact that the equation of state is given by expression $\rho = \rho(p)$, *i.e.* the mass density depends only on pressure, the integrand may be written as the differential of some function. In our case this is the function P, of course: $\frac{1}{\rho} dp = dP$ and, therefore,

$$\frac{1}{\rho} \nabla p = \nabla P. \tag{9.11}$$

Now the equation of motion in the Lamb form (9.6) is as follows

$$\partial_t \vec{v} + \nabla \left(\frac{1}{2} |\vec{v}|^2 + \Phi + P \right) = \vec{v} \times \vec{\omega}. \tag{9.12}$$

With this particular case of the equations of motion, we shall analyze the assumption of barotropy of continuum. In the beginning we consider a simpler version of the stationary flow.

9.3.1. Stationary Flow; the Bernoulli Equation

If at each point **x** of the space occupied by the body $\mathcal{B}$, characteristics of the motion do not change over time, the movement (or flow) is called stationary or steady. Mathematically, this means that partial derivatives with respect to time in the equations describing the flow are equal to zero:

$$\partial_t \vec{v} = 0, \quad \partial_t \rho = 0, \quad \partial_t \varepsilon = 0.$$

The equation of motion (9.6) then takes the form

$$\nabla \left(\frac{1}{2}|\vec{v}|^2 + \Phi \right) = -\frac{1}{\rho}\nabla p + \vec{v} \times \vec{\omega}. \tag{9.13}$$

If, in addition, the fluid is barotropic, then by virtue of (9.11) we have

$$\nabla \left(\frac{1}{2}|\vec{v}|^2 + P + \Phi \right) = \vec{v} \times \vec{\omega}. \tag{9.14}$$

We now recall the Euler's formula (3.10) and write the expression for the total time derivative of a function f. In the stationary case it takes the form $d_t f = (\vec{v}, \nabla f)$. Comparing this expression with (9.14), we see that in order to find the total time derivative of the expression in parentheses in the left-hand side, it suffice to calculate the scalar product of (9.14) and $\vec{v}$:

$$d_t \left(\frac{1}{2}|\vec{v}|^2 + P + \Phi \right) = (\vec{v}, \vec{v} \times \vec{\omega}) = 0. \tag{9.15}$$

The right-hand side of (9.15) is evidently equal to zero, because the vector product is orthogonal to each of its factors, and hence the second factor of the scalar product is orthogonal to $\vec{v}$. Thus, in the steady flow the rate of change of $H \equiv \frac{1}{2}|\vec{v}|^2 + P + \Phi$ is equal to zero and this quantity at a point of a body does not change,

$$H = const, \tag{9.16}$$

i.e. H conserves while a point is moving along its trajectory. Equation (9.16) is called the *Bernoulli equation*[3], and the quantity H is called the *Bernoulli integral*.

At arbitrary points of the space occupied by the body, this function, according to (9.14), is defined by expression

$$\nabla H = \vec{v} \times \vec{\omega}$$

and is generally non-uniform. But it is easy to see that *for irrotational flow* ($\vec{\omega} = 0$) *the function H is spatially uniform*, because in this case the gradient $\nabla H = 0$. The above statement is known as the *Bernoulli's theorem*.

As a simple example of usage of the Bernoulli equation, consider the steady flow of a barotropic fluid in a gravitational field of the Earth. If the axis $x_3 = z$ is directed upward from the Earth's surface, the density of the body force is determined by the vector $\vec{b} = (0, 0, -g)$. The potential of the body force (here, usually referred to as *geopotential*) will be equal to $\Phi = gz$ and the Bernoulli equation takes the form:

$$H = \tfrac{1}{2}|\vec{v}|^2 + P + gz. \qquad (9.17)$$

This equation implies that for a given value of z, say, $z = 0$, the value of P (and if $\rho = const$ then just the pressure p) is maximal at points where $\vec{v} = 0$. Such a point usually exists on the surface of the body streamlined by a flow and is called the critical point (see. Fig. **9.1**).

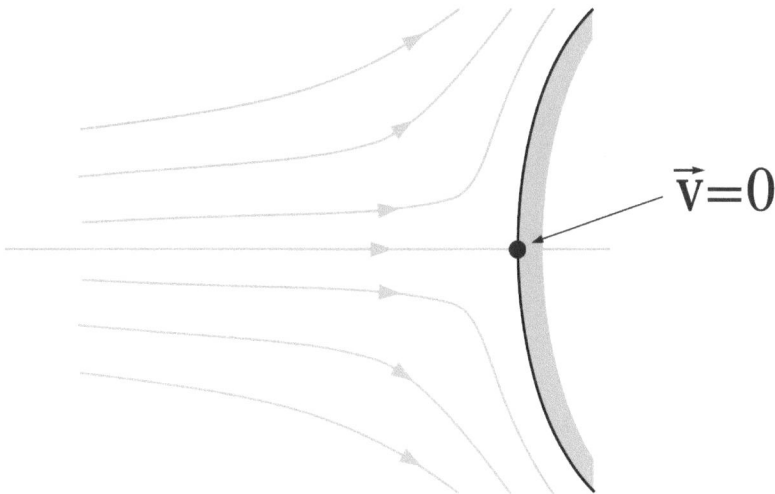

Fig. (9.1). A critical point on the surface of a body streamlined by a stationary fluid flow.

If far from the body the pressure and velocity are equal to p_∞ and $\vec{v}_\infty$, respectively, the pressure at the critical point in case of $\rho = const$ is easily may be found from (9.17) and is equal to

$$p = p_\infty + \frac{1}{2}\rho|\vec{v}_\infty|^2.$$

9.3.2. Trajectories and Streamlines

As noted above, the function H is constant along trajectories of points of the fluid, moving stationary. The same conclusion is sometimes formulated differently, as constancy of the Bernoulli integral along the so-called streamlines. Consider this concept.

Trajectories of points in the general case are described by the equation

$$d_t\mathbf{x} = \vec{v}(t, \mathbf{x}) \qquad\qquad\qquad \textbf{(9.18)}$$

and are those curves, tangent vector to which at every point $\mathbf{x}$ at time t is equal to $\vec{v}(t, \mathbf{x})$. Any particular trajectory is determined by corresponding initial condition

$$\mathbf{x}|_{t=0} = \mathbf{X}.$$

The assumption of stationarity of the flow means that the velocity field is independent of time, and, hence, the equation (9.18) should be replaced by the equation $d_t\mathbf{x} = \vec{v}(\mathbf{x})$.

We can formulate the related problem. Let's fix a velocity field at some time instant t_* and draw a set of curves tangent to vectors of the vector field at each point. In order to write this analytically, we assume that the sought curves are parameterized by some real parameter, denoted, for example, by λ. Thus, the points of these curves are numbered, and the number of a point is the value of λ. Then calculating the derivative $d_\lambda\mathbf{x}$ of position of the point on the curve with respect to parameter λ we obtain the tangent vector to the curve at the given point. Equating the tangent vectors to vectors of the field, we obtain the desired equation

$$d_\lambda\mathbf{x} = \vec{v}(t_*, \mathbf{x}), \qquad\qquad\qquad \textbf{(9.19)}$$

which holds at time t_*. Integral curves of the derived equation (9.19) are called *streamlines*. Any particular streamline is determined by initial condition

$$\mathbf{x}|_{\lambda=0} = \mathbf{x}_0.$$

Compare this equation with (9.18), where the derivative is calculated with respect to t (time), which is the parameter of the world-lines and trajectories (the projections of the world-lines).

The difference between the two totalities of curves, trajectories and streamlines, is as follows. The streamlines correspond to the velocity field associated with a fixed time instant (for example, a snapshot of a velocity field), while trajectories are connected with live, changing over time, velocity field. They are projections of world lines, and therefore keep the knowledge of all past and all future instantaneous vector fields.

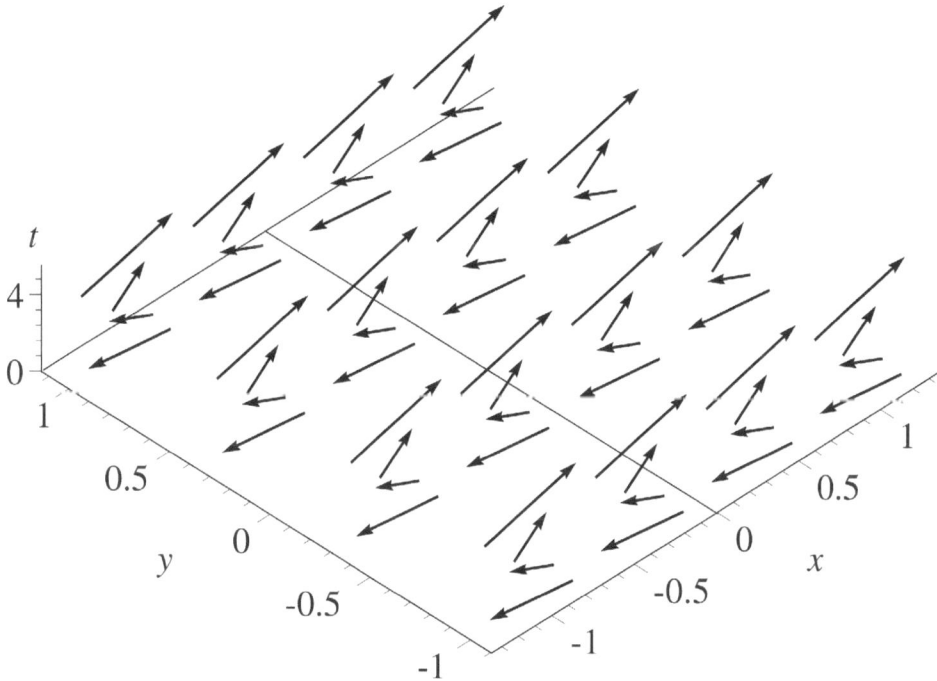

Fig. (9.2). Vector field, tangent to world lines.

Let us illustrate this by the example of a non-stationary vector field governed by the system of equations

$$d_t x = 2(t - a),$$
$$d_t y = b. \tag{9.20}$$

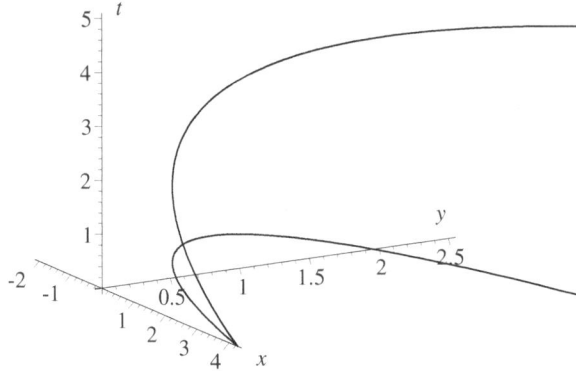

Fig. (9.3). A world-line (upper curve) and corresponding trajectory.

To be able to draw vectors of this vector field we consider the two-dimensional space of places. Here, x, y are the Cartesian coordinates of a point, a, b are some numeric parameters, t is the time. Vectors $\vec{u}$ which are tangent to world lines, have following components $(1, 2(ta), b)$ with respect to coordinate basis. This vector field is shown in Fig. **9.2** for four time instants. Our field is arranged such that at any given time instant all vectors of the field may be obtained from each other by parallel translation. Their projections onto the space of places, *i.e.* vectors $\vec{v}$, which are tangent to trajectories, have components $(2(ta), b)$ with respect to the same basis.

An example of the world-line corresponding to the field $\vec{u}$ and its projection onto the space of places (*i.e.*, the trajectory of the same point of the body with tangent vector $\vec{v}$) are shown in Fig. **9.3**. Both curves are parabolas. The events, which form the world-lines have the coordinates $(t,\ x = (ta)^2 + c_1,\ y = bt + c_2)$. Coordinates of points of the trajectories are $(x = (ta)^2 + c_1,\ y = bt + c_2)$, respectively. Constants c_1 and c_2 are arbitrary. In the given example these coefficients are $a = 2.5$, $b = 0.5$, $c_1 = -2$ and $c_2 = 0$. Thus, the set of trajectories is a family of parabolas.

The set of streamlines, on the contrary, at any time is a totality of parallel lines. Indeed, the system of equations describing the flow lines at time t_*, is

$$d_\lambda x = 2(t_* - a),$$
$$d_\lambda y = b.$$

The variable λ is a parameter of a streamline which numbers its points. The solution of this system are following functions

$t = 0.5$

$t = 2.5$

$t = 5.5$

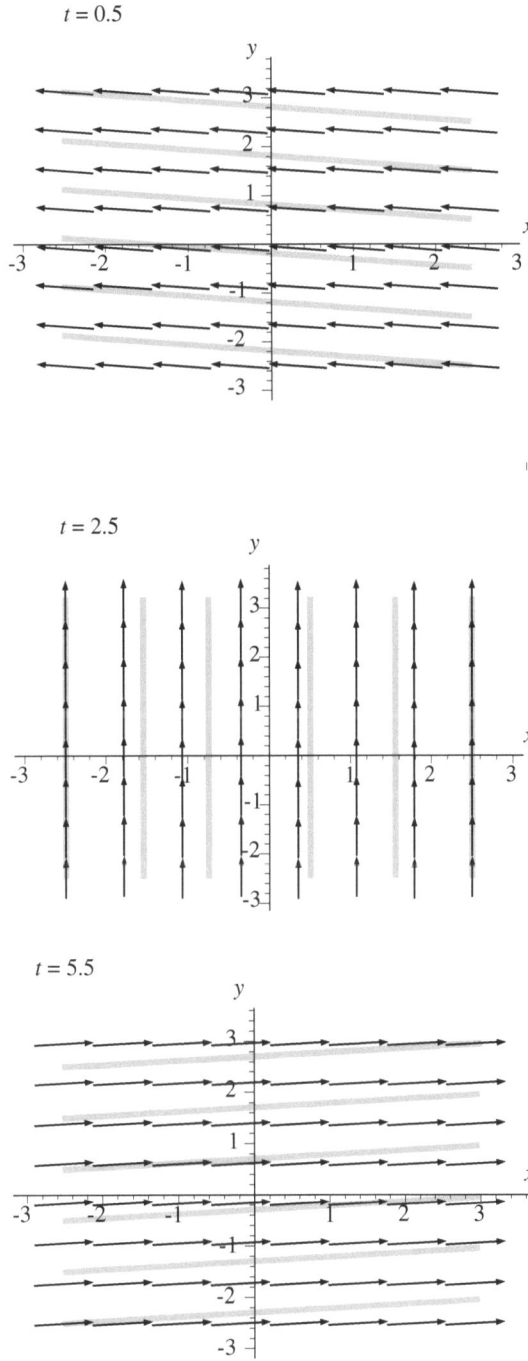

Fig. (9.4). Streamlines and corresponding tangent vector fields at different time instants.

$$x = 2(t_* - a)\lambda + c_1,$$
$$y = b\lambda + c_2.$$

Here c_1 and c_2 are constants of integration. Excluding variable λ, we obtain the equation of the line

$$y = \frac{b}{2(t_*-a)}(x - c_1) + c_2 = \frac{b}{2(t_*-a)}x + const.$$

Vector fields and corresponding streamlines for three successive time instants are given in Fig. **9.4**. It is easy to see that in case of a nonstationary vector field trajectories and streamlines differ from each other.

If the flow is stationary, the velocity field at any time is the same $\vec{v} = \vec{v}(\mathbf{x})$. Neither in past vector fields nor in future, there is nothing different from present ones, and therefore trajectories coincide with streamlines. Now we are studying just such flows. They say that in this case the equation (9.16) is valid along the streamline, *i.e.* the function H is constant along this set of points.

9.3.3. Nonstationary Flow; the Vorticity Equation

The assumption of barotropy also allows significant simplification of the equation of motion, in general nonstationary case. Simplification is based on the fact that the curl of the gradient of any twice differentiable function f is equal to zero:

$$\nabla \times \nabla f = 0. \tag{9.21}$$

Consider again the equation of motion (9.12)

$$\partial_t \vec{v} + \nabla \left(\frac{1}{2}|\vec{v}|^2 + \Phi + P \right) = \vec{v} \times \vec{\omega}.$$

Taking the curl of both sides of this equation, by virtue of (9.21), we obtain

$$\nabla \times (\partial_t \vec{v}) = \nabla \times (\vec{v} \times \vec{\omega}).$$

Since the order of partial differentiation in the mixed derivative is indifferent, the curl and time derivative may be interchanged, and then the expression in the left-hand side will be as follows $\partial_t(\nabla \times \vec{v}) = \partial_t \vec{\omega}$, and the equation itself takes the form

$$\partial_t \vec{\omega} = \nabla \times (\vec{v} \times \vec{\omega}). \tag{9.22}$$

This equation contains only one unknown quantity, the velocity (either explicitly or as the curl $\vec{\omega}$), and hence, it may be solved independently of the rest equations of the model. This result, greatly simplifies the problem. The derived equation describes evolution of the vortex field $\vec{\omega}$ and is therefore called the *vorticity equation·*

NOTES

[1]Contain time derivatives of the 1st order only.

[2]Lamb, Horace (1849-1934), a British mathematician and engineer.

[3]Bernoulli Daniel (1700–1782), a Swiss mathematician.

Classical Fluid Mechanics, 2017, 123-134

CHAPTER 10

Incompressible Perfect Fluid

Abstract: A model of the perfect fluid with an additional assumption of incompressibility of the medium is considered. We derive the Helmholtz equation and formulate the Lagrange and Helmholtz theorems which follow from this equation. We also discuss two-dimensional flows and introduce a new concept of stream function. Next we consider potential flows and study the relationship between the velocity potential and the stream function for a stationary two-dimensional flow.

Keywords: Barotropic model, Helmholtz equation, Helmholtz theorem, Incompressible perfect fluid, Lagrange theorem, Potential flow, Rotational flow, Stream function, Velocity potential, Vortex line.

Experimental data show that the dropping liquids are extremely poorly compressible objects. Gases at sufficiently low speeds behave in the same way. All these observations allow using in some cases as an additional hypothesis the assumption of incompressibility of the medium, *i.e.*, to consider that relations (6.20) hold. In this chapter we discuss some of the consequences of the adoption of such a hypothesis. Model of the incompressible perfect fluid is a modified version of the system of equations (9.1 - 9.4):

$$(\nabla, \vec{v}) = 0,$$
$$\partial_t \vec{v} + (\vec{v}, \nabla)\vec{v} = -\frac{1}{\rho}\nabla p + \vec{b},$$
$$d_t \varepsilon = (\nabla, \vec{h}) + \rho s,$$
$$\rho = \rho(p, \varepsilon).$$

Assumption of incompressibility $d_t \rho = 0$ affects only the first (the continuity equation) and the third (the internal energy balance) equations of the model. In the absence of outer heating ($\vec{h} = 0$ and $s = 0$) the internal energy conserves.

10.1. BAROTROPIC MODEL

The assumption of the barotropy allowed reducing the equation of motion to the vorticity equation. Let's find out what new brings the adoption of the incompressibility hypothesis to the model.

10.1.1. The Helmholtz Equation

Important results about the motion of incompressible perfect fluid may be obtained from consideration of the vorticity equation. For this purpose, we transform the right-hand side of (9.22) using the known formula from the vector analysis, which is valid for two arbitrary vector fields[1]:

$$\nabla \times (\vec{v} \times \vec{\omega}) = \vec{v}(\nabla, \vec{\omega}) - \vec{\omega}(\nabla, \vec{v}) + (\vec{\omega}, \nabla)\vec{v} - (\vec{v}, \nabla)\vec{\omega}. \tag{10.1}$$

The first two terms on the right-hand side of this expression are zero. Firstly, we consider an incompressible fluid, and hence, $(\nabla, \vec{v}) = 0$. Secondly,

$$(\nabla, \vec{\omega}) = (\nabla, \nabla \times \vec{v}),$$

and it is known that for any differentiable vector field $\vec{f}$ the equality

$$(\nabla, \nabla \times \vec{f}) = 0$$

holds. Substituting the remaining two terms in (9.22), we obtain:

$$\partial_t \vec{\omega} + (\vec{v}, \nabla)\vec{\omega} = (\vec{\omega}, \nabla)\vec{v}. \tag{10.2}$$

The expression on the left-hand side is the total time derivative of the vorticity vector. Writing it on the left side, we finally obtain the so-called *Helmholtz*[2] *equation*:

$$d_t \vec{\omega} = (\vec{\omega}, \nabla)\vec{v}. \tag{10.3}$$

This equation may be solved. At first, we rewrite it in an equivalent form, taking into account that $(\vec{\omega}, \nabla)\vec{v} = G\vec{\omega}$ where $G \equiv \nabla \vec{v}$:

$$d_t \vec{\omega} = G\vec{\omega}. \tag{10.4}$$

Now recall that, discussing the kinematics of deformation, we have expressed the velocity gradient G in terms of the gradient of deformation F (see formula (5.9) on p.69):

$$G = (d_t F)F^{-1}.$$

Substituting this expression into the equation (10.4), one finds:

$$d_t \vec{\omega} = (d_t F)F^{-1}\vec{\omega}. \tag{10.5}$$

Now we rewrite the left-hand side of (10.5) in the form

$$d_t \vec{\omega} = d_t(I\vec{\omega}),$$

represent the unit tensor as a contraction $I = FF^{-1}$ and perform differentiation

$$d_t(I\vec{\omega}) = d_t(FF^{-1}\vec{\omega}) = (d_t F)F^{-1}\vec{\omega} + Fd_t(F^{-1}\vec{\omega}). \tag{10.6}$$

Comparing (10.5) and (10.6), we find that

$$Fd_t(F^{-1}\vec{\omega}) = 0. \tag{10.7}$$

This expression may be regarded as an equation for the unknown vector $d_t(F^{-1}\vec{\omega})$, *i.e.*, as a system of linear algebraic equations with respect to unknown components of this vector. Since we assume that motion is reversible[3], the matrix of the tensor F is non-singular, and hence the equation (10.7) has only the trivial solution:

$$d_t(F^{-1}\vec{\omega}) = 0. \tag{10.8}$$

Now, we shall formulate the Cauchy problem for this equation. Assume that at time $t = 0$ in the region considered, a vector field $\vec{\omega}_0$ is defined. At the same time instant the deformation is absent and $F = F^{-1} = I$. Hence, the initial condition for the equation (10.8) takes the form

$$F^{-1}\vec{\omega}|_{t=0} = \vec{\omega}_0. \tag{10.9}$$

Integrating the Cauchy problem (10.8 - 10.9), we obtain:

$$F^{-1}\vec{\omega} = \vec{\omega}_0 \qquad \Rightarrow \qquad \vec{\omega} = F\vec{\omega}_0. \tag{10.10}$$

This result implies important consequences, known as the Lagrange and Helmholtz theorems.

10.1.2. The Lagrange and Helmholtz Theorems

Recall that the flow is called irrotational if the vorticity is zero everywhere.

Theorem (Lagrange). *The (ir)rotational mode of the flow of the perfect incompressible barotropic fluid is conserved.*

Indeed, if the flow was irrotational at the initial time instant, then $\vec{\omega}_0 = 0$ everywhere. According to (10.10) at any other time instant the vorticity vector $\vec{\omega}$ at each point will also be zero.

Conversely, let at time $t = 0$ the flow is rotational, *i.e.*, there exists at least one point where $\vec{\omega}_0 \neq 0$. Then, since the deformation gradient F is not equal to the zero tensor, the vorticity vector $\vec{\omega} = F\vec{\omega}_0$ at this point will also be different from the zero vector at any other time instant.

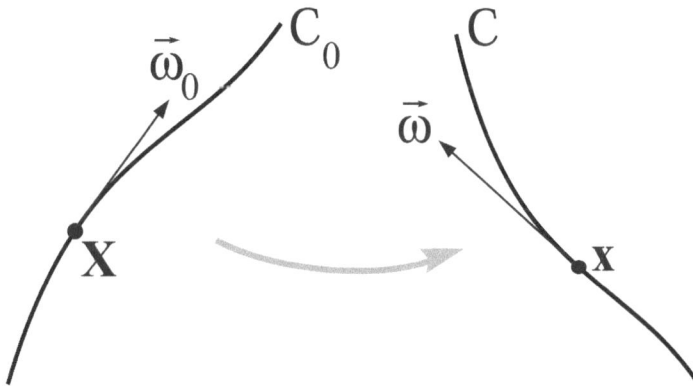

Fig. (10.1). Deformation of a vortex line.

Now, before formulating the Helmholtz theorem we introduce some definitions. A curve which at each point is tangent to a continuous vector field is called a *vector line*. If the vector field is a velocity field corresponding to a certain time instant, the vector lines are called *streamlines*. They have already been discussed on p.117.

If the vector field is a vorticity field, the vector lines are called *vortex lines*.

Vector lines passing through a closed loop, which can be contracted to a point, form a surface called a *vector tube*. In case of streamlines, such a tube is called the *stream tube*, and in case of the vortex lines, the *vortex tube*, respectively.

Theorem (Helmholtz). *Vortex lines in the incompressible, barotropic, perfect fluid are conserved.* In other words, *a set of points of the body, constituting a vortex line at the initial time instant, is the vortex line at any other time* also.

Consider in the initial time $t = 0$ a vortex line C_0 (see Fig. **10.1**), parameterized by a real parameter λ. Let a vector $d_\lambda \mathbf{X} = \vec{\omega}_0$ tangent to C_0 is defined at the point with coordinates $\mathbf{x}|_{t=0} = \mathbf{X}$. Due to deformation of the vortex line C_0 at time t it is mapped to some line C. Let's show that this new curve will also be a vortex line.

By virtue of the deformation a point $\mathbf{X}$ is mapped to a point $\mathbf{x}$ and the vector tangent to the curve C at this point is determined by the expression $d_\lambda \mathbf{x}$. In accordance with the formula (5.1), this quantity is equal to

$$d_\lambda \mathbf{x} = F d_\lambda \mathbf{X} = F d_\lambda C_0 = F \vec{\omega}_0.$$

Considering now the formula (10.10), we obtain:

$$d_\lambda \mathbf{x} = \vec{\omega},$$

and this means that the vector tangent to the curve C is the vorticity vector. Similar arguments are valid for any point of each vortex line C_0. Hence, vortex lines transform into vortex lines, which proves the Helmholtz theorem.

10.2. TWO-DIMENSIONAL FLOW

The description of physical phenomena may be significantly simplified in cases when the full three-dimensional system of equations may be reduced to a system of smaller dimension. For example, if there is a direction in which velocity does not change, it is possible to choose a coordinate system such that two components of the velocity of a fluid will depend only on two coordinates, and the third component will be equal to zero

$$\vec{v} = \vec{v}(x_1, x_2) = (v_1, v_2, 0) \equiv (u, v, 0).$$

In this two-dimensional (flat) case the streamlines are closely related to another, often used element of description of the flows, the stream function.

10.2.1. Stream Function

In two dimensions the continuity equation of the incompressible fluid takes the form:

$$(\nabla, \vec{v}) = \partial_x u + \partial_y v = 0, \qquad (10.11)$$

or

$$\partial_x u = -\partial_y v.$$

Let's introduce a function $\psi(x, y)$ with mixed derivative, such that

$$\partial_y \psi = u, \qquad \partial_x \psi = -v. \qquad (10.12)$$

Then the equation (10.11) is automatically satisfied:

$$\partial_{xy} \psi - \partial_{yx} \psi \equiv 0.$$

The function ψ is called the *stream function*. The equation which solution is the stream function, is derived from the equation of motion of incompressible perfect fluid, or rather, from the Helmholtz equation.

Let's calculate the vorticity in the considered flat case:

$$\vec{\omega} = (\partial_{x_2} v_3 - \partial_{x_3} v_2, \partial_{x_3} v_1 - \partial_{x_1} v_3, \partial_{x_1} v_2 - \partial_{x_2} v_1) =$$
$$= (0, 0, \partial_x v - \partial_y u) = \omega_z \vec{e}_3,$$

where
$$\omega_z \equiv \partial_x v - \partial_y u = -\partial_{xx} \psi - \partial_{yy} \psi = -\Delta \psi.$$

Here Δ denotes the second order differential operator $\Delta \equiv \partial_{xx} + \partial_{yy}$, which is called the two-dimensional Laplace operator[4]. Now we substitute the found expression for vorticity in the Helmholtz equation

$$\partial_t \vec{\omega} + (\vec{v}, \nabla)\vec{\omega} = (\vec{\omega}, \nabla)\vec{v},$$

calculating scalar products beforehand. Taking into account that $\vec{\omega} = -\Delta \psi \vec{e}_3$ for the term $(\vec{v}, \nabla)\vec{\omega}$ we have

$$(\vec{v}, \nabla)\vec{\omega} = -(\partial_y\psi\,\partial_x - \partial_x\psi\,\partial_y)\Delta\psi\vec{e}_3 = \left(-\partial_y\psi\,\partial_x\Delta\psi + \partial_x\psi\,\partial_y\Delta\psi\right)\vec{e}_3.$$

Since the velocity components are functions of two arguments x and y, we obtain

$$(\vec{\omega}, \nabla)\vec{v} = \omega_z\,\partial_z\vec{v} = 0.$$

The Helmholtz equation now looks as follows

$$\partial_t\Delta\psi + \partial_y\psi\,\partial_x\Delta\psi - \partial_x\psi\,\partial_y\Delta\psi = 0,$$

or in the stationary case

$$\partial_y\psi\,\partial_x\Delta\psi - \partial_x\psi\,\partial_y\Delta\psi = 0.$$

Previously obtained the vorticity equation is a substantial simplification of the problem, because it allows one to find the velocity *via* solving the single vector equation. In the current two-dimensional case we are able to go further and reduce the problem for the velocity of the flow to a single scalar equation for the stream function.

10.2.2. Streamlines of Stationary Flow

Knowing the stream function, it is possible to directly determine the form of streamlines for stationary flows. Indeed, since the rate of change of the stream function along some world line is equal to

$$d_t\psi = \partial_t\psi + u\,\partial_x\psi + v\,\partial_y\psi,$$

then in the stationary case we have

$$d_t\psi = u\,\partial_x\psi + v\,\partial_y\psi = 0. \tag{10.13}$$

The last equality follows from the definition of the stream function (10.12). In the stationary flow the stream function is constant along streamlines, since in this case they coincide with trajectories.

If we imagine the stream function as the surface $\psi(x, y)$, then streamlines will be level lines of this surface and will be determined by the equation

$$\psi(x, y) = C = const.$$

10.2.3. Fluid Flow Through a Contour

Let's apply the notion of stream function to calculation of the fluid flow through a liquid contour and to determination of dependence of the flow on the form of this contour.

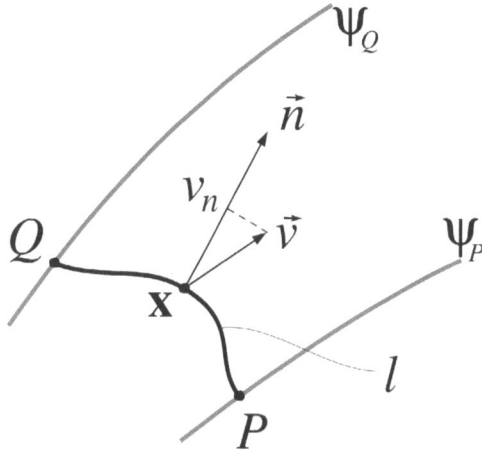

Fig. (10.2). Fluid flow through a contour l .

Consider a contour l bounded by two points P and Q (see Fig. **10.2**). A flow q of the fluid with constant mass density ρ through this contour, by definition, is equal to

$$q = \rho \int_P^Q v_n dl = \rho \int_P^Q (\vec{v}, \vec{n}) dl. \tag{10.14}$$

Here $\vec{v}$ is the velocity of the flow at point **x** on the curve l, $\vec{n}$ is the unit normal to the curve at the same point, and v_n is the projection of $\vec{v}$ on the direction of the normal. In order to calculate q we must know the normal vector $\vec{n}$. Let us make up a system of equations with respect to the unknown components of $\vec{n}$.

Let $\vec{dl}$ be a tangent vector to the contour l at the point **x**. Then $(\vec{n}, \vec{dl}) = 0$ is the first equation of the required system. Next, if dx and dy are components of the tangent vector with respect to the Cartesian basis, *i.e.*, $\vec{dl} = (dx, dy)$, then the square of its length dl^2 is equal to the scalar product $(\vec{dl}, \vec{dl})$. This gives the second equation of the system. Finally, taking into account the fact that $\vec{n}$ is the unit vector, *i.e.*, $(\vec{n},\ \vec{n}) = 1$, we obtain following system of equations with respect to the unknown components n_x and n_y of the normal vector:

$$n_x dx + n_y dy = 0, \tag{10.15}$$

$$dx^2 + dy^2 = dl^2, \tag{10.16}$$

$$n_x^2 + n_y^2 = 1. \tag{10.17}$$

This system may be reduced to a simpler linear system for the unknowns. Multiplying the last two equations and subtracting the square of the first one, we obtain:

$$(n_x dy - n_y dx)^2 = dl^2.$$

Extracting the square root from both sides of this equality, we find the relation which together with equation (10.15) gives the necessary system:

$$n_x dx + n_y dy = 0,$$
$$n_x dy - n_y dx = \pm dl.$$

Solving it, we find $\quad n_x = \pm\dfrac{dy}{dl}, n_y = \mp\dfrac{dx}{dl}$. Let's choose the direction of the normal such that the scalar product $(\vec{v},\ \vec{n})$ is positive. Then $n_x = \dfrac{dy}{dl}, n_y = -\dfrac{dx}{dl}$ and $(\vec{v}, \vec{n}) = u\dfrac{dy}{dl} - v\dfrac{dx}{dl}$ Substituting the scalar product found in the integral (10.14), we obtain:

$$q = \rho \int_P^Q (u\,dy - v\,dx) = \rho \int_P^Q (\partial_x \psi\,dx + \partial_y \psi\,dy) = \rho \int_P^Q d\psi = \rho(\psi_Q - \psi_P),$$

where ψ_Q and ψ_P are values of the stream function on the streamlines passing through corresponding points. Thus, the mass flow through the contour is determined by the difference of values of the stream function at the ends of the contour. The form of the contour does not affect the value of the flow.

The flat problem considered may be generalized to three dimensions. A contour between two streamlines is an analogue of a three-dimensional cross-section of the stream tube. The flow of the fluid through any cross-section of the considered stream tube must be the same.

10.3. THE POTENTIAL FLOW

If the motion is irrotational, then $\vec{\omega} = 0$, *i.e.*, the curl of velocity vector $\nabla \times \vec{v}$ is identically zero everywhere in the domain, occupied by the flow. Since it is known that for arbitrary twice differentiable function the gradient of its curl is identical zero, the velocity vector of irrotational flow may be written as the gradient of a scalar quantity φ

$$\vec{v} = \nabla \varphi. \tag{10.18}$$

The above-introduced function φ is called the *velocity potential* and an irrotational flow is also called a *potential flow*, *i.e.*, a flow, where velocity possesses a potential[5].

The equation that describes evolution of the velocity potential may be obtained from the continuity equation. Indeed, substituting the definition of potential in the continuity equation for incompressible fluid (10.18), we find:

$$(\nabla, \vec{v}) = (\nabla, \nabla \varphi) = \Delta \varphi = 0, \tag{10.19}$$

where $\Delta = \partial_{xx} + \partial_{yy} + \partial_{zz}$ is the three-dimensional Laplace operator. To determine the velocity potential the Neumann problem is posed, *i.e.*, the boundary condition:

$$\partial_n \varphi = v_n \tag{10.20}$$

is set up on the boundary of the region under consideration. Here v_n is a normal component of velocity at the boundary and ∂_n is a derivative with respect to the normal to the boundary. The problem (10.19), (10.20) allows one to find potential up to an additive constant. Since the velocity potential is a solution of the Laplace equation, it satisfies the *maximum principle*: the maximal absolute values of the function φ are located on the boundary of the domain occupied by the flow.

Some conclusions about the nature of the flow may be made based on the type of the problem (10.19), (10.20), which determines the velocity potential and, therefore, the velocity field itself. Thus, if the potential flow is generated by a solid body moving in the fluid, the characteristics of the flow at any time instant depend only on the instantaneous speed of the body but not on its acceleration. This is due to the fact that the Laplace equation does not contain time derivatives, and the boundary condition contains only the speed of the body.

10.4. RELATIONSHIP BETWEEN VELOCITY POTENTIAL AND STREAM FUNCTION

If the potential flow is flat, we may introduce a stream function which also satisfies the Laplace equation. Indeed, the expression $\Delta\psi = -\omega_z$ connecting the component of vorticity with the stream function , was obtained on p.128. Meanwhile, since in the present case, the vorticity is equal to zero, we have $\Delta\psi=0$

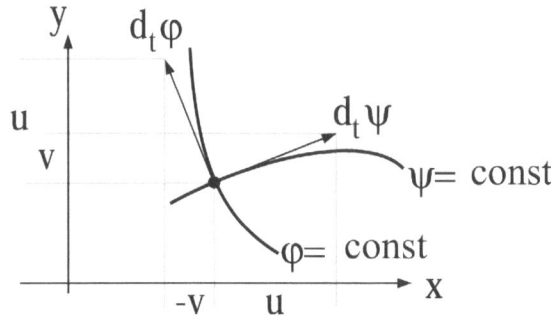

Fig. (10.3). Connection between the potential φ and the stream function ψ.

However, the relationship between the potential and the stream function, is closer. In order to detect it, we express the velocity components in two ways: in terms of the stream function according to formulas (10.12), and *via* the velocity potential using the formula (10.18):

$$u = \partial_y\psi = \partial_x\varphi, \qquad v = -\partial_x\psi = \partial_y\varphi.$$

Substituting these relations into the expression (10.13) which holds in the steady case along the streamline, we find:

$$d_t\psi = u\,\partial_x\psi + v\,\partial_y\psi = (\partial_x\varphi)\,\partial_x\psi + (\partial_y\varphi)\,\partial_y\psi = (\nabla\varphi, \nabla\psi) = 0. \textbf{(10.21)}$$

The quantities $u=\partial_x\varphi$ and $v=\partial_y\varphi$ are components of the vector tangent to the curve $\psi=$ const. In turn, the quantities $\partial_x\psi, \partial_y\psi$ may be considered as components of the vector tangent to the curve $\varphi=$ const. Indeed, from (10.21) and the stream function definition we have

$$(\nabla\varphi, \nabla\psi) = (\partial_x\psi)\,\partial_x\varphi + (\partial_y\psi)\,\partial_y\varphi = (-v)\,\partial_x\varphi + u\,\partial_y\varphi = d_t\varphi = 0.$$

The vector $\nabla\psi$ with components $(-v, u)$ is tangent to the line $\varphi=$ const at the point where it is defined. Tangent vectors $\nabla\psi$ and $\nabla\psi$ according to (10.21) are orthogonal (see, Fig. **10.3**). Thus, the equality (10.21), which is the scalar product of tangent vectors may be interpreted as orthogonality of the curves $\psi =$ const and $\varphi =$ const. From the geometric viewpoint it does not matter what is considered a streamline or an equipotential curve. If some hydrodynamic problem is solved, *i.e.* the characteristics of a flow defined by the functions φ and ψ are found, at the same time a related problem, where ψ plays the role of the velocity potential, and φ is the stream function, is also solved. Such problems as well as functions ψ and φ are called *conjugate*.

NOTES

[1] Note that, while summation or multiplication of vectors, like any other algebraic operation, make sense for arbitrary vectors of some vector space, differential operations require the presence of a vector field.

[2] Helmholtz Hermann Ludwig Ferdinand (1821–1894), a German physicist, mathematician, physiologist and psychologist.

[3] You should understand that this assumption is a description of limits of applicability of the theory. In case the motion of the fluid cannot be considered reversible, one should be prepared for the fact that this theory fails to operate, *i.e.*, will describe and predict the behavior of the fluid motion with errors exceeding the permissible level.

[4] Laplace Pierre Simon (1749–1827) is a French astronomer, mathematician and physicist.

[5] The situation here is just the same as in the case of the body force density. The quantity $\vec{b}$ in most geophysical problems may be written as the gradient of geopotential precisely because according to the observational data, the field $\vec{b}$ is irrotational with great accuracy.

Classical Fluid Mechanics, 2017, 135-144

CHAPTER 11

Viscous Fluid

Abstract: We study the modification of the stress tensor to account for shear stresses, and derive the Navier-Stokes equation of motion of the viscous fluid. Next we discuss the problem posing for the viscous fluid model, and also consider the energy balance together with the viscous dissipation of the kinetic energy.

Keywords: Dissipation of the kinetic energy, Dynamic viscosity, Energy balance, Internal friction, Navier-Stokes equation, No-slip condition, Slip condition, Viscous fluid, Viscous stress tensor.

11.1. PECULIARITIES OF THE PERFECT FLUID MODEL

The choice of the stress tensor in the form (7.24) is good in the sense that corresponding model of the perfect fluid may, in some cases successfully, describe flows of a real fluid. However, there are situations when this does not happen, and predictions made by this model strikingly differ from observations. Apparently, accepting model of the perfect fluid, we neglect, among other things, those properties of real media that in these situations become significant. Thus, for example, the definition (7.24) implies that there are no shear stresses in a model continuum (see (7.25)). It is clear that in cases where they are small compared to normal, they probably may be neglected. But where this is not the case, the model of the perfect fluid will not work.

What causes the absence of shear stresses? The stress is the density of contact forces of interaction of a body with environment. For example, if a fluid is moving in a channel, the flow interacts with walls of the channel. If they solely compress the fluid only normal stresses are non-zero. However, if there is a fluid friction against the walls of the channel, the shear stresses are non-zero too. Assuming the absence of shear stresses, we neglect the friction of the fluid on the walls of the channel. Besides, there is no friction between layers of fluid along any kinematic boundary. On the one hand this leads to the fact that such a model of continuum turns out to be almost insensitive to introduction of any objects into the flow domain. And on the other hand provides preservation in time of rotational or potential mode of the flow. The absence of friction prevents generation of vorticity, resulting in the absence of mixing.

11.2. THE EQUATION OF MOTION OF THE VISCOUS FLUID

Let's try to take into account the shear stresses using more complicated construction of the stress tensor T. However we shall not change it drastically (because in some cases the existing tensor works well), but shall simply add a correction term T′:

$$T = -p\mathrm{I} + T'. \tag{11.1}$$

This additional term just like the first one, is a tensor of the rank 2. Its task is to approximately describe forces of inter-molecular interactions obstructing the existence and development of inhomogeneities, which are characterized by non-zero gradients. These forces will be called viscous forces, and the tensor T′ in turn will be called the viscous stress tensor.

From experience we know that the forces of viscosity do not show up when the fluid moves as a rigid body, without stretching or compression. This flow regime is characterized by the deformation rate tensor equal to zero D = O (see. p.72). In all other cases, when D ≠ O, viscous forces smooth down the velocity field and tend to reduce the flow of the fluid into the motion of the rigid body. Viscous forces behave this way in any fluid. For this reason, it is natural to suppose that T′ depends on D and to consider the simplest dependence T′ ∼ D. Such dependence is also reasonable because the tensor D is the symmetric part of the velocity gradient $D = \frac{1}{2}(G + G^{\mathrm{T}})$ and the viscous stress tensor T′ must be symmetric because of the symmetry of the stress tensor T.

Different substances possess different ability to smooth out heterogeneities and the viscous stress tensor must take this into account. This may be most easily achieved by the introduction of the proportionality factor, whose value reflects the specific character of each individual substance. With this regard we set by definition

$$T' = 2\mu D. \tag{11.2}$$

The quantity μ is called the *dynamic viscosity*. It is always positive, since the tensor T′ always acts the same way. This is an experimental fact. Deuce in (11.2) is set for convenience, because the definition of the tensor D contains coefficient

$\frac{1}{2}$ and we get rid of it in this way (surely, it might be possible to omit this deuce, but then we would have to double the numerical value of μ). So, the new stress tensor is as follows

$$\mathrm{T} = -p\mathrm{I} + 2\mu\mathrm{D} = -p\mathrm{I} + \mu(\mathrm{G} + \mathrm{G}^\mathrm{T}), \tag{11.3}$$

and its components are equal to

$$T_{ij} = -p\delta_{ij} + \mu\big(\nabla_j v_i + \nabla_i v_j\big). \tag{11.4}$$

Before we substitute the expression (11.3) in the Cauchy law of motion (7.22), we calculate the divergence of the new stress tensor:

$$div\mathrm{T} = div(-p\mathrm{I} + 2\mu\mathrm{D}), \tag{11.5}$$

Its *i*-th component is equal to:

$$\begin{aligned}
(div\mathrm{T})_i &= \nabla_j\Big(-p\delta_{ij} + \mu\big(\nabla_j v_i + \nabla_i v_j\big)\Big) = \\
&= -\nabla_i p + \nabla_j\big(\mu\nabla_j v_i\big) + \nabla_j\big(\mu\nabla_i v_j\big).
\end{aligned} \tag{11.6}$$

To demonstrate the effect of the individual terms, we introduce additional simplifying assumptions. Assume that

1. $\mu = $ const.
2. the coordinate system is Cartesian and $\nabla_k = \partial_{x_k}$, consequently.

Now the *i*-th component of divergence of the stress tensor is as follows

$$\begin{aligned}
(div\mathrm{T})_i &= -\partial_{x_i}p + \mu\Delta v_i + \mu\,\partial_{x_j x_i}v_j = \\
&= -\partial_{x_i}p + \mu\Delta v_i + \mu\,\partial_{x_i}(\nabla, \vec{v}).
\end{aligned} \tag{11.7}$$

In the Cartesian coordinates the Laplacian Δ takes the form: $\Delta = \partial_{x_i x_i}$. If non-Cartesian coordinates are selected, the expression for Δ is more complicated (non-zero Christoffel symbols should be taken into account) and for a number of known cases it may be found in the reference books.

Substituting (11.5) into the law of motion (7.22), we obtain the so-called *Navier-Stokes-Duhem*[1] *equation of motion of the viscous fluid*:

$$\rho d_t \vec{v} = -\nabla p + 2\mu div D + \rho \vec{b}. \tag{11.8}$$

As already noted, the actual fluid in most cases may be considered incompressible medium. This property is often immediately taken into account in the model, which allows some simplification of the equation of motion. Let's substitute the divergence of the stress tensor in the form (11.7) into the law of motion (7.22)

$$\rho d_t \vec{v} = -\nabla p + \mu \Delta \vec{v} + \mu \nabla (\nabla, \vec{v}) + \rho \vec{b}. \tag{11.9}$$

Since the continuity equation in this case has the form ($\nabla, \vec{v}$) = 0 the third term in the right-hand side of (11.9) is zero and the equation takes the form of the *Navier-Stokes equation*

$$\rho d_t \vec{v} = -\nabla p + \mu \Delta \vec{v} + \rho \vec{b}. \tag{11.10}$$

11.3. THE EQUATION OF MOTION OF THE VISCOUS FLUID IN COMPONENT FORM

The vector equation of motion of the viscous fluid, similar to the Euler's equation (7.28), may be written in the component form as three scalar equations. We choose the Cartesian coordinate system for simplicity, and construct the coordinate basis $\{\vec{e}_i\}$ (see p.92 for details) at each point of the space of places. The equation (11.9) we write as follows

$$\partial_t \vec{v} + (\vec{v}, \nabla)\vec{v} + \frac{1}{\rho}\nabla p - \nu \Delta \vec{v} - \nu \nabla(\nabla, \vec{v}) - \vec{b} = 0, \tag{11.11}$$

where $\nu \equiv \frac{\mu}{\rho}$ is the so-called *kinematic viscosity*. Using previously written expressions (7.29) for components of the vectors, we get:

$$\left(\partial_t v_i + (\vec{v}, \nabla)v_i + \frac{1}{\rho}\nabla_i p - \nu \Delta v_i - \nu \nabla_i(\nabla, \vec{v}) + g\delta_{3i}\right)\vec{e}_i = 0, \tag{11.12}$$

and further

$$\partial_t v_i + (\vec{v}, \nabla)v_i = -\frac{1}{\rho}\nabla_i p + \nu \Delta v_i + \nu \nabla_i(\nabla, \vec{v}) - g\delta_{3i}. \tag{11.13}$$

Now substituting the component form of the scalar product $(\vec{v}, \nabla) = v_k \nabla_k$, the divergence of velocity $(\nabla, \vec{v}) = \nabla_k v_k$ and the Laplace operator $\Delta \equiv (\nabla, \nabla) = \nabla_k^2$, we finally obtain

$$\partial_t v_i + v_k \nabla_k v_i = -\frac{1}{\rho} \nabla_i p + \nu \nabla_k^2 v_i + \nu \nabla_i (\nabla_k v_k) - g \delta_{3i}. \qquad (11.14)$$

In case of incompressible fluid (the Navier-Stokes equation) instead of (11.14), we obtain a simpler equation:

$$\partial_t v_i + v_k \nabla_k v_i = -\frac{1}{\rho} \nabla_i p + \nu \nabla_k^2 v_i - g \delta_{3i}. \qquad (11.15)$$

Note that the free index i numbers the equations, and k is a dummy index.

11.4. PROBLEM FORMULATION IN FLUID MECHANICS OF THE VISCOUS FLUID

In contrast to the model of the perfect fluid the system of equations of the viscous fluid includes second-order derivatives of components of the velocity vector with respect to spatial variables. For this reason, the well-posedness of the problem requires additional boundary conditions, which are usually set along all borders of the region occupied by the flow. From physical considerations the *slip condition* at the boundary of the perfect medium (the values of tangential component of velocity of the fluid and velocity of the boundary are independent) for the viscous fluid is replaced by the *no-slip* or *adhesion condition*. In this case, the velocity of the fluid on an impermeable boundary is equal to velocity of the boundary. In addition, it is assumed that when crossing the kinematic border, the stress vector $\vec{\tau}$ changes continuously.

11.5. VISCOUS DISSIPATION AND THE ENERGY BALANCE

Since we have changed our view on the contact forces acting on the surface of a configuration of a body or any part thereof, and this has reflected in the structure of the stress tensor, it is obvious that the power of the contact forces and its distribution has changed as well. According to (8.7 , 8.10 and 8.11) the power of the contact forces is as follows

$$W_C = \int_\chi (\vec{v}, div\mathrm{T}) dV + \int_\chi (\mathrm{T} : \nabla \vec{v}) dV, \qquad (11.16)$$

where both terms describe the rate of doing work by contact forces: changing the kinetic energy of the body (the first term) and changing its internal energy (the second term).

Substituting the stress tensor in the form (11.3) into (11.16) we find [2]:

$$W_C = \int_\chi \left(-(\vec{v}, \nabla p) + 2\mu(\vec{v}, div\mathrm{D}) \right) dV + \int_\chi \left(-p(\nabla, \vec{v}) + 2\mu \mathrm{D} : \nabla \vec{v} \right) dV. \quad \textbf{(11.17)}$$

First term in each integral still describes the working of the pressure forces, while the second term describes the work of the viscous forces. At the same time, the contribution of shear stresses in the evolution of the internal energy (the last term in the second integral) has an important property: it is non-negative.

To demonstrate this property we show that it may be written in the form of a square of a certain quantity. Firstly, we write the inner product $2\mathrm{D} : \nabla \vec{v}$ in component form:

$$\left(\partial_{x_k} v_j + \partial_{x_j} v_k \right) \partial_{x_k} v_j = \left(\partial_{x_k} v_j \right) \left(\partial_{x_k} v_j \right) + \left(\partial_{x_j} v_k \right) \left(\partial_{x_k} v_j \right).$$

In these expressions all indices are dummy and may be denote by any letter. Working with indices we shall try to write down the inner product as a square of the sum of components of the velocity gradient. The first term is the square $\left(\partial_{x_k} v_j \right)^2$. We divide it into two equal terms, and in one of them change the designation of indices: the index k we shall denote by letter j and the former index j by letter k. Then we obtain the expression which is just the required square

$$\frac{1}{2} \left(\partial_{x_k} v_j \right)^2 + \frac{1}{2} \left(\partial_{x_j} v_k \right)^2 + \left(\partial_{x_k} v_j \right) \left(\partial_{x_j} v_k \right) = \frac{1}{2} \left(\partial_{x_k} v_j + \partial_{x_j} v_k \right)^2.$$

Hence,

$$\left(\partial_{x_k} v_j + \partial_{x_j} v_k \right) \partial_{x_k} v_j = \frac{1}{2} \left(\partial_{x_k} v_j + \partial_{x_j} v_k \right)^2 = 2\mathrm{D} : \mathrm{D}. \quad \textbf{(11.18)}$$

Such a result reflects the general property of symmetric tensors: the inner product of symmetric tensor with antisymmetric one is equal to zero. Hence, the inner product of a symmetric tensor with an arbitrary tensor is equal to the inner product of a symmetric tensor with the symmetric part of the second tensor. Thus,

$$\text{D}:\nabla\vec{v} = \text{D}:\text{D} \geq 0 \quad \Rightarrow \quad 2\mu\int_\chi \text{D}:\nabla\vec{v}dV = 2\mu\int_\chi \text{D}:\text{D}dV \geq 0, \qquad \textbf{(11.19)}$$

and, indeed, this part of the power of the contact forces is never less than zero. But what's the big deal?

The important fact is that, according to our results, the work of shear stresses leads only to the increase of the internal energy. According to the first law of thermodynamics (8.21) the internal energy may decrease, either as a result of expansion of the body, or due to the negative heating rate, *i.e.* the heat removal by thermal conductivity in contact with a colder body, or by radiation.

Consider the case of a thermally insulated body, when the first law of thermodynamics is given by the equation (8.20). Using the representation of the power of forces acting on the body (8.17), as well as definitions of the potential and internal energies (p.101 and beyond), we write the energy balance (8.20) in the form

$$d_t((K + U) + E) = \int_\chi div(\text{T}\vec{v})dV. \qquad \textbf{(11.20)}$$

Here on the right side is the power of the contact forces, which determines the rate of change of the sum of mechanical $(K + U)$ and internal E energies. At that, the vanishing of the contact forces on the surface of thermally insulated body

$$d_t((K + U) + E) = 0 \qquad \textbf{(11.21)}$$

still does not mean constancy of certain types of energy and lack of redistribution of energy between its various components. So, the internal energy will not increase only when the contact forces are zero at any kinematic boundary within the body. In particular, this means that D = 0 at each point of the configuration, and the fluid moves as a rigid body without stretching or compression. The inequality D ≠ 0 valid at least at one point means the increase of internal energy, which continues, until the mode of motion of a rigid body is set in.

Let the potential energy of the fluid is constant. Then the conservation law of the total energy implies that an increase in the internal energy of the body $E(\mathcal{B})$ is due to the simultaneous reduction of its kinetic energy $K(\mathcal{B})$. This process, as well as the quantity

$$Diss \equiv 2\mu \int_{\chi} D:D dV, \tag{11.22}$$

is called the *dissipation of kinetic energy*. The dissipation manifests itself everywhere where the deformation rate is non-zero, and there are movements of fluid particles relative to each other. Since such movements are accompanied by the alignment of the velocity, *i.e.* the inhibition of fast particles of the fluid and the acceleration of slow particles by friction, it is said that the dissipation of kinetic energy is due to forces of the *internal friction*.

Differential balance equations of kinetic and internal energy densities may be easily obtained from (8.6 and 8.31) using corresponding substitution of the stress tensor (11.3):

$$\rho d_t \frac{k}{\rho} = -(\vec{v}, \nabla p) + 2\mu(\vec{v}, div D) + (\vec{v}, \rho \vec{b}), \tag{11.23}$$

$$\rho d_t \frac{\varepsilon}{\rho} = -p(\nabla, \vec{v}) + 2\mu D:D + (\nabla, \vec{h}) + \rho s. \tag{11.24}$$

To demonstrate the exchange of energy between kinetic and internal components of the total energy, we exclude from consideration the body forces (in the first equation) and external heating (in the second equation). In addition, in the first two terms of the right side of (11.23), we isolate the divergences of the vectors $p\vec{v}$ and $D\vec{v}$, respectively, while the left side of (11.23) we transform using the continuity equation[3]

$$\rho d_t \frac{k}{\rho} = \rho \, \partial_t \frac{k}{\rho} + (\rho \vec{v}, \nabla \frac{k}{\rho}) + \frac{k}{\rho} \underbrace{(\partial_t \rho + (\nabla, \rho \vec{v}))}_{=0} =$$

$$= \left(\rho \, \partial_t \frac{k}{\rho} + \frac{k}{\rho} \partial_t \rho\right) + \left((\rho \vec{v}, \nabla \frac{k}{\rho}) + \frac{k}{\rho}(\nabla, \rho \vec{v})\right) =$$

$$= \partial_t k + (\nabla, k\vec{v}).$$

As a result, we find

$$\partial_t k + (\nabla, k\vec{v}) = -(\nabla, p\vec{v}) + p(\nabla, \vec{v}) + 2\mu(\nabla, (D\vec{v})) - 2\mu D:D,$$

or, after combining divergence in both sides of the equation,

$$\partial_t k + (\nabla, (k+p)\vec{v} - 2\mu D\vec{v}) = p(\nabla, \vec{v}) - 2\mu D:D. \tag{11.25}$$

In turn the equation (11.24) may be written as:

$$\partial_t \varepsilon + (\nabla, \varepsilon \vec{v}) = -p(\nabla, \vec{v}) + 2\mu \mathrm{D}: \mathrm{D}. \qquad \textbf{(11.26)}$$

Integrating both equations over a fixed volume V we obtain the integral balance relations:

$$\partial_t K + \int_V (\nabla, (k+p)\vec{v} - 2\mu \mathrm{D}\vec{v})dV = \int_V p(\nabla, \vec{v})dV - Diss,$$
$$\partial_t E + \int_V (\nabla, \varepsilon \vec{v})dV = -\int_V p(\nabla, \vec{v})dV + Diss.$$

Now we transform the integrals on the left-hand sides using the Gauss' theorem:

$$\partial_t K + \int_S \left((k+p)\vec{v} - 2\mu \mathrm{D}\vec{v}\right)_n dS = \int_V p(\nabla, \vec{v})dV - Diss,$$
$$\partial_t E + \int_S (\varepsilon \vec{v})_n dS = -\int_V p(\nabla, \vec{v})dV + Diss.$$

Here index n labels normal components of the flux densities of the corresponding integral parameters K and E. Let's finally assume that the flow through the surface S of the volume is equal to zero, then the left-hand sides of the balance equations will contain only one term:

$$\partial_t K = \int_V p(\nabla, \vec{v})dV - Diss, \qquad \textbf{(11.27)}$$

$$\partial_t E = -\int_V p(\nabla, \vec{v})dV + Diss. \qquad \textbf{(11.28)}$$

Now it is possible to draw conclusions.

Firstly, we see that the sum of the two equations gives the conservation of the total energy in the absence of external inflows: the right sides cancel each other out.

Secondly, the increase in the kinetic energy means the decrease of the internal energy, and *vice versa*. This may be due to expansion or compression of the body by pressure.

And finally, thirdly, regardless of the work of the pressure forces there is a sink of the kinetic energy and, consequently, the source of the internal energy. This is the dissipation. Unlike pressure forces which may be both the source and the sink of both types of energy the dissipation is always unidirectional. It always increases the internal energy at the expense of the kinetic energy and never *vice versa*[4].

NOTES

[1]Navier (Navier) Louis Marie Henri (1785–1836), a French physicist and mathematician. Duhem (Duhem) Pierre Maurice Marie (1861–1916), a French physicist, philosopher and historian of science.

[2]Here, as well as above, we are using the simplifying assumptions 1 and 2 introduced on p.137.

[3]Here we utilize a particular case of the widely used method of converting expressions like $\rho d_t a$ where a is a differentiable function. Using the Euler's formula, the total derivative is written as

$$\rho d_t a = \rho\, \partial_t a + \rho v_k \nabla_k a.$$

Summing further, this expression with the continuity equation, multiplied by a, we obtain

$$\rho d_t a = \rho\, \partial_t a + \rho v_k \nabla_k a + a\, \underbrace{(\partial_t \rho + \nabla_k(\rho v_k))}_{=0} =$$
$$= (\rho\, \partial_t a + a\, \partial_t \rho) + (\rho v_k \nabla_k a + a \nabla_k(\rho v_k)) =$$
$$= \partial_t(\rho a) + \nabla_k(\rho a v_k).$$

[4]And just in case, don't forget that this is a model of a natural phenomenon, but not the nature itself. In the nature, perhaps, not all so straightforward.

CHAPTER 12

Related Topics

Abstract: The issues connected with material of the previous chapters are considered. We define the temperature and derive the heat equation. Next we discuss the free convection and obtain the system of equations describing this phenomenon. We study the applicability of the perfect and viscous fluid models, introduce the non-dimensional form of equations of the fluid model, and consider the dynamic similarity of the flows.

Keywords: Dynamic similarity, Fourier law of heat conduction, Free convection, Froude number, Heat equation, Non-dimensional equations, Reynolds number, Temperature, Thermal conductivity, Thermal diffusivity.

12.1. THE HEAT EQUATION

If the internal energy is a quantity more convenient to construct a theory, in practice it is often preferable to work with another quantity, the temperature, which is easier to measure. Consider the case of an ideal gas when the *temperature* of the medium T is determined by the following relation:

$$d_t \frac{\varepsilon+p}{\rho} = c_p d_t T, \tag{12.1}$$

where c_p is the heat capacity at constant pressure, and the quantity $\frac{\varepsilon+p}{\rho}$ is called the specific *enthalpy*.

The evolution equation for the temperature may be derived from the internal energy balance equation (8.32) for the perfect fluid or (11.24) for the viscous fluid, using the substitution $d_t \frac{\varepsilon}{\rho} = c_p d_t T - d_t \frac{p}{\rho}$ obtained from the definition (12.1). In case of the perfect fluid we get:

$$\rho c_p d_t T = d_t p + (\nabla, \vec{h}) + \rho s. \tag{12.2}$$

Next, it is required to set the heat flux density vector $\vec{h}$ on the boundary of configuration. An experimental fact is that the hotter body transfers heat to less heated. Mathematically, the direction of the greatest change of temperature is described by its gradient, therefore quite reasonable is the assumption of proportionality of the vector $\vec{h}$ to the temperature gradient ∇T

$$\vec{h} = \lambda \nabla T. \tag{12.3}$$

This relationship is called the *Fourier law of heat conduction*. The function of temperature λ is called *thermal conductivity*. Its task is to describe the ability of a particular substance to conduct heat. Combining further, (12.2) with (12.3), we obtain the heat equation in the moving continuum:

$$\rho c_p d_t T = d_t p + (\nabla, \lambda \nabla T) + \rho s. \tag{12.4}$$

The function λ is often assumed constant and is taken outside the scalar product. Besides, if we take into account that the operator $(\nabla, \nabla)(\,\cdot\,) = \Delta(\,\cdot\,)$ is the *Laplace operator*, it is possible to write

$$\rho c_p d_t T = d_t p + \lambda \Delta T + \rho s. \tag{12.5}$$

Usually, the first term on the right-hand side $(d_t p)$ is small (due to two assumptions: the local thermodynamic equilibrium $d_t \frac{p}{\rho} = 0$ and incompressibility $d_t \rho = 0$) and is neglected. In this case, after normalization by ρc_p the equation (12.5) is written in the form

$$d_t T = \partial_t T + (\vec{v}, \nabla T) = \kappa \Delta T + s_T, \tag{12.6}$$

where $\kappa \equiv \dfrac{\lambda}{\rho c_p}$ is the so-called *thermal diffusivity* and $s_T \equiv \dfrac{s}{c_p}$ is the specific power of the volumetric heat source (due to radiation, chemical reactions, and so on). This equation is applicable to fluids, and gases, if the velocity of the latter is much less than the speed of sound. If problems of fluid mechanics are formulated in terms of temperature, and the heat equation (12.4) is used instead of the internal energy balance equation, it is more convenient to write the equation of state in terms of temperature, rather than the internal energy density, *i.e.* to search it in the form $\rho = \rho(p, T)$[1].

Formally the equation (12.6) coincides with the *diffusion equation*

$$d_t C = \partial_t C + (\vec{v}, \nabla C) = \kappa \Delta C + s_C, \qquad (12.7)$$

which describes evolution of a passive admixture concentration $C(t, \mathbf{x})$ in the medium. Coefficient κ in this case is called the coefficient of *molecular diffusion* and s_C is the specific power of the admixture source.

The admixture is called *passive* if it does not affect the flow dynamics. This assumption means that the velocity field $\vec{v}$ may be determined regardless of C and then used in the diffusion equation for calculation of the concentration field C.

If we consider the temperature as a passive admixture we, thus, assume that the temperature inhomogeneities are so small that they do not affect the dynamics. The velocity field may be calculated without using the internal energy balance equation (or heat equation) and the equation of state takes the form (8.36). In this case, temperature inhomogeneities only shuffled over and smoothed due to molecular heat conductivity. Such a motion of non-uniformly heated fluid is often called the *forced convection*. In contrast, the *free convection* is an example of a flow where the temperature cannot be considered a passive admixture.

12.2. FREE CONVECTION

Consider the motion of a non-uniformly heated fluid under the influence of buoyancy forces that cause sinking of a cold fluid mass and lifting of a heated one. This motion is called the *free convection*. The temperature cannot be considered here as a passive admixture, because its heterogeneities generate motion themselves. We derive the equations describing fluid motion in this mode.

Let's assume that the mass density is dependent on temperature only, and the fluid is incompressible (*i.e.*, the continuity equation is written in the form (6.20)). Let the origin of the Cartesian coordinate system is located on the surface of the fluid layer and z-axis is directed upward.

Firstly, we consider the case without horizontal motion, $u = v = 0$. The system of equations of the fluid model in this case may be written as:

continuity equation: $\quad \partial_z w = 0,$

equations of motion: $\quad \partial_x p = 0,$

$$\partial_y p = 0, \tag{12.8}$$

$$\partial_t w = -\frac{1}{\rho}\partial_z p - g,$$

heat equation: $\quad \partial_t T = \kappa \Delta T,$

equation of state: $\quad \rho = \rho(T).$

If we assume that the vertical component of velocity w at the fluid surface is equal to zero, then from the continuity equation we obtain $w=0$ everywhere in the fluid layer.

If the fluid temperature is constant $T = T_0 =$ const, then the equation of state gives $\rho(T_0) = \rho_0 =$ const and the system (12.8) transforms into the already considered system of equations of hydrostatics. Integrating the equations of motion for $p|_{z=0} = p_* =$ const, we find

$$p|_{T=T_0} = p_0(z) = p_* - \rho_0 g z. \tag{12.9}$$

Suppose now that the temperature is non-constant. Let it be written in the form $T(t, \mathbf{x}) = T_0 + T_1(t, \mathbf{x})$, where $T_0 =$ const. Then from the equation of state up to terms of the first order of smallness we obtain

$$\begin{aligned}
\rho(t, \mathbf{x}) = \rho(T(t, \mathbf{x})) &= \rho(T_0) + (T - T_0)(\partial_T \rho)_{T=T_0} + o(T - T_0) \tag{12.10}\\
&\approx \rho_0 + T_1(\partial_T \rho)_{T=T_0}\\
&= \rho_0 - T_1 \beta \rho_0\\
&= \rho_0 + \rho_1(t, \mathbf{x}),
\end{aligned}$$

where $\rho_1(t, \mathbf{x}) = -\rho_0 \beta T_1(t, \mathbf{x})$ is the change in the mass density of a medium under the heating, and $\beta = -\frac{1}{\rho_0}(\partial_T \rho_0)_{T=T_0}$ is the so-called *coefficient of thermal expansion*. The pressure is given by (12.9) and in this case is determined by the expression

$$p(t, \mathbf{x}) = p_* - \rho(t, \mathbf{x})gz = (p_0 + \rho_0 gz) - \rho gz \qquad (12.11)$$
$$= p_0 - (\rho - \rho_0)gz$$
$$= p_0 - \rho_1 gz$$
$$= p_0 + p_1.$$

Here p_1 is the pressure change due to heating of the medium. Using these formulas we transform the right side of the equation of motion for the vertical component of velocity w. We write the approximate expression for $\frac{1}{\rho} \partial_z p$ at the point (ρ_0, p_0) using the Taylor series expansion of $\frac{1}{\rho}$ with respect to ρ_0:

$$\frac{1}{\rho} = \frac{1}{\rho_0} - \frac{(\rho - \rho_0)}{\rho_0^2} + \cdots = \frac{1}{\rho_0} - \frac{\rho_1}{\rho_0^2} + \cdots .$$

Then

$$\frac{1}{\rho} \partial_z p = \left(\frac{1}{\rho_0} - \frac{\rho_1}{\rho_0^2} + \cdots \right) \partial_z (p_0 + p_1)$$
$$\approx \frac{1}{\rho_0} \partial_z p_0 + \frac{1}{\rho_0} \partial_z p_1 - \frac{\rho_1}{\rho_0^2} \partial_z p_0.$$

Only linear terms with respect to ρ_1 and p_1 are withheld in the latter expression. Since $\partial_z p_0 = -\rho_0 g$ in the first approximation we have

$$\frac{1}{\rho} \partial_z p \approx -g + \frac{1}{\rho_0} \partial_z p_1 + \frac{\rho_1}{\rho_0} g = -g + \frac{1}{\rho_0} \partial_z p_1 - g\beta T_1.$$

Substituting this result into (12.8), we obtain

$$\partial_t w = -\frac{1}{\rho_0} \partial_z p_1 + g\beta T_1.$$

Thus, the temperature (as a deviation from the mean) is contained directly in the equation of motion and is either a source of this movement or an additional stabilizing factor.

In the more general case of motion with velocities low enough to make it possible to neglect the influence of the mass density changes due to changes in pressure, it is easy to write a system of equations that describes the free convection in fluid. It is sufficiently to bring back all derivatives, which currently are equal to zero. This system is as follows

$$(\nabla, \vec{v}) = 0,$$

$$d_t u = -\frac{1}{\rho_0} \partial_x p,$$

$$d_t v = -\frac{1}{\rho_0} \partial_y p,$$

$$d_t w = -\frac{1}{\rho_0} \partial_z p_1 + g\beta T_1,$$

$$d_t T = \kappa \Delta T,$$

$$\rho = \rho(T).$$

12.3. WHAT FLUID IS VISCOUS?

Now we have two versions of the fluid model, corresponding to different definitions of the stress tensor:

1. The model of the perfect fluid with following equation of motion

$$d_t \vec{v} = -\frac{1}{\rho} \nabla p + \vec{b}, \tag{12.12}$$

2. The model of the viscous fluid with modified (regarding the previous case) equation of motion

$$d_t \vec{v} = -\frac{1}{\rho} \nabla p + \vec{b} + 2\nu div\mathrm{D}. \tag{12.13}$$

We proceed from the assumption that all natural fluids are viscous. Does this mean that after the model of the viscous fluid is developed, the initial variant is hopelessly outdated and may be discarded?

Why? The model of the perfect fluid, indeed, does not describe the interaction of molecules with each other. But, firstly, it was not envisaged at creation of the model and therefore is rather its property, not its defect. Secondly, all models (the model of the viscous fluid including) do not describe something, precisely because they are models, and not the nature itself. Moreover, when creating a model we deliberately seek to take into account not all possible details, since the task of researchers and, especially, of engineers, is not to copy the nature, but to obtain an efficient method for solving problems, which include rather a limited range of characteristics of a phenomenon. Thus, estimating a particular model, we should consider not only its ability to reproduce various aspects of a phenomenon, but also how effectively it allows us to do this. It is clear that if, say, a model of the weather will issue forecasts with the rate of change of the weather itself, they will hardly be needed, even if they would be very accurate.

All the above fully applies to both our models. The accounting of any additional property of a real fluid is not given for nothing. Therefore, where the effect of viscosity may be neglected, probably it would be more reasonable to use a simpler model of the perfect fluid. And let us not forget that in any case a real fluid is not at all what it appears in the model.

The initial question obviously must be formulated differently. For example, this way: in what case the description of the moving medium should be based on the model of the perfect (viscous) fluid? Let's try to answer.

It easily can be seen that the perfect fluid approximation gives good results where the viscous forces are small compared with the inertia forces. Both forces are represented by corresponding terms in the equation of motion. In order to assess the impact of various terms, the equations are written down in the so-called *non-dimensional form*.

12.3.1. Equations in the Non-dimensional Form

Physical quantities are usually expressed by numbers obtained *via* comparison with standards, which are called the *basic units of measurement* (for example, foot, ton, barrel). A set of basic units, sufficient to describe a given class of phenomena is called the *system of units* (most frequently SI, CGS, MKS are used). A multitude of derivative units are formed from the basic units. The dimension of a physical quantity is called a function that determines in how many times the numerical value of this quantity will change under the transition from the original system of units to another. It is a power monomial that depends on the main units only (for example, $kg^{\alpha} \cdot m^{\beta} \cdot s^{\gamma}$ or $kW^{\delta} \cdot h^{\varepsilon}$. Here $\alpha, \dots, \varepsilon$ are integer constants). To denote the dimension of a certain quantity a, the square brackets are used $[a]$. The dimension of non-dimensional quantity is equal to unity. A certain quantity a_1 of a set $(a_1, \dots, a_n)$ has an independent dimension if this dimension cannot be represented as a product of powers of dimensions of the other quantities $(a_2, \ldots, a_n)$. For example, among three quantities ρ, p and $\vec{v}$, only any two of them have independent dimension. Indeed, the dimensions of these quantities in the SI system are

$$[\rho] = kg \cdot m^{-3}, \quad [p] = N \cdot m^{-2} = kg \cdot m^{-1} \cdot s^{-2}, \quad [\vec{v}] = m \cdot s^{-1},$$

and there exist integer constants a and b such that $[p] = [\rho]^a \cdot [\vec{v}]^b$. These constants are easy to find and they are equal to $a = 1, b = 2$. On the contrary, the quantities $\rho, \mathbf{f}_{\mathcal{B}}$ and $\vec{v}$ all have independent dimensions, *i.e.* integer constants a and b such that $[\mathbf{f}_{\mathcal{B}}] = [\rho]^a \cdot [\vec{v}]^b$ do not exist (see, *e.g.*, [7] for details).

In order to obtain the non-dimensional form of equations, each dimensional variable of the problem a_d (dependent and independent variables, but not parameters) is expressed as a product of its typical value A and a non-dimensional function a:

$$a_d(t, \mathbf{x}) = A\, a(t, \mathbf{x}). \tag{12.14}$$

Each constant A obviously has the same dimension as the variable a_d and is called its *scale*. It may be either an average value of a or an average value of its amplitude. It is chosen such that the non-dimensional variable a would be of order of unity within the problem under study. The number of independent scales, *i.e.* scales with independent dimensions must be equal to the number of independent dimensions.

When all dimensional variables are written in the form (12.14), each term of each equation is represented as a product of a dimensional factor (which consists of the chosen scales and parameters of equations), non-dimensional variables and their derivatives (also non-dimensional). Since each scale is chosen such that corresponding non-dimensional variable is of order of unity, the value of a dimensional factor may be considered as an estimate of contribution of corresponding term into the equation.

If we normalize the equation by any of these dimensional factors, the entire equation turns out to be non-dimensional due to its uniformity in dimension, and the values of non-dimensional coefficients in different terms of equation will represent their relative contributions.

Let's demonstrate this procedure on the example of the Navier-Stokes equation (11.10). We introduce the dimensional scales for existing variables which have the following mutually independent dimensions:

$$[t] = s, \qquad [\mathbf{x}] = m, \qquad [p] = \frac{kg}{m \cdot s^2}.$$

Thus, there should be three independent scales. In order to exclude the dimension of mass we normalize the equation by ρ. Now it contains the ratio $\frac{1}{\rho} \nabla p$ with dimension m/s^2. Hence, there are only two independent dimensions and only two independent scales are necessary. All other scales would be derivatives from these two.

We choose the velocity scale U and the length scale L (not specifying their numerical values) as independent scales[2]. Then the ratio L/U turns to be the time scale T. We write dimensional variables of the problem, denoted herein by the index d , in the form (12.14)

$$\vec{v}_{\mathrm{d}} = U\vec{v}, \quad t_{\mathrm{d}} = Tt, \quad \partial_{t_{\mathrm{d}}} = \frac{U}{L}\partial_t,$$

$$\left(\frac{1}{\rho}\nabla p\right)_{\mathrm{d}} = \frac{U^2}{L}\frac{1}{\rho}\nabla p, \quad (\Delta\vec{v})_{\mathrm{d}} = \frac{U}{L^2}\Delta\vec{v}.$$

Please note, that differential operators in physical sciences are usually also dimensional. Substituting these expressions into the equation (11.10), we obtain:

$$\frac{U^2}{L}\partial_t\vec{v} + \frac{U^2}{L}(\vec{v},\nabla)\vec{v} = -\frac{U^2}{L}\frac{1}{\rho}\nabla p + \nu\frac{U}{L^2}\Delta\vec{v} + \vec{b}. \qquad \textbf{(12.15)}$$

This equation is still dimensional like the initial equation (11.10). We only have separated the dimensional quantities (here they are the factors $\frac{U^2}{L}$, $\nu\frac{U}{L^2}$ which are formed by the scales and parameters) and non-dimensional variables. Normalizing the equation by one of these factors we obtain the non-dimensional equation of motion. Since three terms in the equation (249) have the same factor $\frac{U^2}{L}$, we choose it for normalization and get the required result:

$$\partial_t\vec{v} + (\vec{v},\nabla)\vec{v} = -\frac{1}{\rho}\nabla p + \left(\frac{\nu}{UL}\right)\Delta\vec{v} + \left(\frac{gL}{U^2}\right)\frac{\vec{b}}{g}.$$

Here all terms are non-dimensional. The relative contribution of the last two terms on the right side is determined by non-dimensional factors in parentheses. Introducing the traditional notation for these factors

$$Re \equiv \frac{UL}{\nu}, \qquad Fr \equiv \frac{U^2}{gL},$$

we finally obtain

$$\partial_t \vec{v} + (\vec{v}, \nabla)\vec{v} = -\frac{1}{\rho}\nabla p + \frac{1}{Re}\Delta\vec{v} + \frac{1}{Fr}\frac{\vec{b}}{g}. \qquad (12.16)$$

The non-dimensional factor Re which is called the *Reynolds*[3] *number* characterizes the ratio of inertial forces to viscous forces, and the non-dimensional factor Fr which is called the *Froude*[4] *number* characterizes the ratio of inertial forces to the gravity force.

12.3.2. The Dynamic Similarity

As may be seen, all the multitude of solutions of (12.16) depends only on two parameters Re and Fr. The quantities U and L characterize spatial and temporal scales of the studied flow. Generally speaking, by specifying different numerical values of U and L we obtain different flows. However, if both non-dimensional numbers Re and Fr do not change their values, various dimensional characteristics of the flows will correspond to one and the same solution of the non-dimensional equation (12.16). In other words, each solution of the non-dimensional equation (12.16) describes a family of flows with the same values of numbers Re and Fr. Besides, if regions occupied by such flows are geometrically similar, flows themselves are called *dynamically similar*.

We emphasize once more the physical meaning of the non-dimensional numbers Re and Fr. Let's write down the characteristic values of the force densities, which determine the fluid flow in terms of selected scales.

1. The density of the inertial force, determined by the product of the mass density and acceleration, is proportional to $\rho U^2/L$.
2. The density of the gravity force is equal to ρg.
3. The density of the friction force generated by viscosity is proportional to $\rho \nu U/L^2$.

The ratio of the inertial force density to the viscous force density is equal to the Reynolds number:

$$\frac{\rho U^2/L}{\rho \nu U/L^2} = \frac{UL}{\nu} \equiv Re.$$

In turn, the ratio of the inertial force density to the gravity force density is equal to the Froude number:

$$\frac{\rho U^2/L}{\rho g} = \frac{U^2}{Lg} \equiv Fr.$$

Consideration of the non-dimensional equation (12.16) shows that for sufficiently large values of Re the term $\frac{1}{Re}\Delta\vec{v}$ becomes negligible, and the flow is well described by the model of the perfect fluid.

Thus, the answer to the question in the headline of the section 4.3 "what fluid is viscous?" is connected not only with the physical properties of the fluid itself, but also with the spatial and temporal characteristics of the flow. Large values of the viscosity ν do not guarantee small values of the ratio $Re = UL/\nu$. The motion of one and the same medium may require either the model of the perfect fluid or the model of the viscous fluid. Everything depends on the space-time scales of the phenomenon under study.

12.4. NON-DIMENSIONAL FORM OF THE HEAT EQUATION

Similar considerations are applicable to the equation (12.6), which describes the heat transfer. The non-dimensional form of the heat equation is easily obtained by sequentially performing the above procedure. The result is as follows:

$$\partial_t T + (\vec{v}, \nabla T) = \frac{1}{Pe}\Delta T. \tag{12.17}$$

Here $Pe = \frac{UL}{\kappa}$ is the *Peclet*[5] *number*, and U and L are the velocity and length scales of the studied flow.

Exercise. Derive the equation (12.17).

By analogy with the Reynolds number, the Peclet number characterizes the ratio of the inertial term ($\vec{v}, \nabla T$) to diffusional $\kappa \Delta T$. If the values of Pe of the flow are large then the evolution of the temperature field is determined mainly by the velocity field $\vec{v}$ and the influence of the molecular heat transfer is negligible. The equation (12.17) takes the form $d_t T = 0$ which means conservation of temperature along the world lines.

Certainly, the equation (12.17) itself does not describe any molecules. After all, we study a continuous medium, smooth temperature field, *etc*. However, since the equation is a model, not nature, and the nature at the microscopic level is discrete, the model has to include mechanism of influence of micro-processes (interaction of molecules) on macro-processes (heat transfer). The simplest and most common way to account for such mechanisms is their parametric description. Here, such a parameter is the coefficient κ. Using different values of this parameter, we try to approximately take into account the ability of molecules (different for different substances) to interact with each other.

NOTES

[1]Along with the heat equation, the so-called entropy balance equation is often considered. The specific *entropy* S is defined similarly to (12.1) by the expression $d_t \frac{\varepsilon + p}{\rho} = T d_t S$, which being substituted in the internal energy balance equation, gives the desired relation. In case of the perfect fluid we find

$$\rho T d_t S = d_t p + (\nabla, \vec{h}) + \rho s.$$

Thus, it follows that if $d_t p = 0$ and the external heating is absent ($\vec{h} = 0, s = 0$) the specific entropy at a point of the perfect fluid is retained. If this equation is used instead of the internal energy balance equation, it makes sense to write the equation of state in the form $\rho = \rho(p, S)$.

[2]This is the traditional choice. Of course, it is possible to select any two other independent scales, for example, L and T. The result will be the same, but will be expressed in other terms.

[3]Reynolds, Osborne (1842-1912), a British physicist and engineer.

[4]Froude, William (1810-1879), a British shipbuilder.

[5]Peclet Jean-Claude (1793-1857), a French physicist.

CHAPTER 13

Turbulent Fluid

Abstract: The modification of the fluid model to describe the turbulent flow is considered. We study the hydrodynamic instability, methods of description of turbulence, the problem of averaging and the energy balance of the turbulent fluid. Finally, we derive the Reynolds equation of motion.

Keywords: Eddy viscosity, Hydrodynamic instability, Laminar mode, Reynolds conditions, Reynolds equation, Reynolds number, Reynolds stress tensor, The averaging problem, Turbulence, Turbulent energy, Turbulent mode.

13.1. THE HYDRODYNAMIC INSTABILITY

Typically, the equations of the viscous fluid model give good description of observational data for sufficiently small values of $Re = \frac{UL}{\nu}$. If the values of Re are large, the solutions of the model equations still exist (in the mathematical sense), but do not match the observed flows. Theoretical solutions describe smooth variations of smooth hydrodynamic fields, whereas random fluctuations in time and space of all hydrodynamic characteristics of the flow are observed.

Observations show that there exist two modes in fluid motions, which differ greatly from each other:

1. *laminar* mode: regular flow, smoothly evolving under the influence of external conditions.

2. *turbulent* mode: chaotic flow, characterized by random spatial-temporal fluctuations of hydrodynamic fields.

Besides, the difference is not only external. Such properties of the flows as the wall friction, mixing rate, thermal conductivity change dramatically. In turbulent flow, all this is much more intense.

General criterion for the onset of turbulence was set by O.Reynolds in the late 19th century. It lies in the fact that the flow is laminar until the number Re does not exceed a certain critical value Re_{cr} and becomes turbulent when $Re > Re_{cr}$.

With increasing Re upon reaching the critical value the perturbations abruptly generate turbulence. The very same value of Re_{cr} depends on intensity and scale of perturbations brought into the flow (*e.g.*, by different obstacles, roughness of walls, *etc.*), *i.e.*, the Reynolds number is not an unambiguous criterion of turbulence. However, it is possible to specify the minimal critical value Re_{crmin} such that for $Re < Re_{crmin}$, the laminar mode of the flow always will recover, *i.e.* any perturbation will decay. The value of Re_{crmin} is about 2000. As for $Re > Re_{crmin}$ the flow becomes unstable with respect to sufficiently large perturbations.

Thus, the number Re_{cr} marks a boundary of stability of both modes, laminar and turbulent. When $Re < Re_{cr}$ the laminar mode is stable and perturbations arising under the influence of inertial forces are suppressed by viscosity. When $Re > Re_{cr}$ conversely, the laminar flow loses stability, viscosity cannot cope with emerging disturbances, they grow and the flow becomes turbulent.

The prevalence of turbulent flows makes the problem of their descriptions particularly important.

13.2. THE DEVELOPED TURBULENCE

Although at subcritical values of Re close to Re_{cr} the flow is still called laminar, the transitional processes occur in the form of brief flashes of high frequency disturbances (the so-called "turbulent spots"). Meanwhile at supercritical values of Re the entire area of the flow is covered with chaotic fluctuations of characteristics of the flow and, in particular, of components of velocity. Trajectories of points of the fluid become extremely mixed up, and what we were trying to avoid by accepting the continuity hypothesis (*i.e.*, the need for individual description of the motion of fluid particles) reappears with the onset of turbulence.

The idea, associated with the acceptance of the continuity hypothesis, consisted in supposition that the pattern of motion of close particles is similar and little changing on small scales. The description of such motion requires relatively small amount of information. The turbulent mode upsets all plans. Initially close trajectories quickly diverge. The amount of information needed to describe these movements increases dramatically, and we essentially find ourselves in the original position. The old methods become unusable for description of turbulent flows and another approach is required, which was found. This is a statistical approach, and corresponding mathematical tool is the theory of random fields.

In the modern statistical fluid mechanics [8] fields of characteristics of a turbulent flow are always assumed to be *random fields*. Each particular realization of such a field is considered as a representative of a *statistical ensemble* of fields. Instead of describing details of individual realizations, averaged fields are being studied.

Any hydrodynamic characteristic f of a turbulent flow is written as the sum of an averaged quantity $\bar{f}$ and deviation from it

$$f = \bar{f} + f'. \tag{13.1}$$

The averaging is understood as the probability-theoretic averaging on the corresponding statistical ensemble. Such averaging, denoted further by the overbar, satisfies the following *Reynolds conditions*:

1. $\bar{a} = a, \quad a = const,$
2. $\overline{af} = a\bar{f}, \quad a = const,$
3. $\overline{\partial_{x_i} f} = \partial_{x_i} \bar{f},$
4. $\overline{f + g} = \bar{f} + \bar{g},$
5. $\overline{\bar{f} g} = \bar{f}\, \bar{g},$ where g can take values $g = 1, \bar{h}$ and h'. In this case, the condition 5) is equivalent to the following three conditions:

$$5a)\ \bar{\bar{f}} = \bar{f}, \quad 5b)\ \overline{\bar{f} \bar{h}} = \bar{f}\, \bar{h}, \quad 5c)\ \overline{h'} = 0.$$

Exercise. Derive conditions 5a)–5c) from the condition 5).

All these conditions actually mean that the averaging operator $\overline{(\cdot)}$ is

▫ Linear: $\overline{af + bh} = a\bar{f} + b\bar{h}, \quad a, b = const,$

▫ Continuous: $\overline{\lim_{n\to\infty} f_n} = \lim_{n\to\infty} \bar{f}_n,$

▫ Projector: $\bar{\bar{f}} = \bar{f}.$

Exercise. Show that conditions 1)–5) follow from the latter three conditions.

13.3. THE AVERAGING PROBLEM

Since in practice it is often impossible to obtain an ensemble of realizations, it is necessary to resort to some other methods of averaging. Let function $f(x)$, which should be smoothed, is defined on the set of real numbers. A function $\bar{f}(x)$ defined by the expression

$$\bar{f}(x) = \int_{-\infty}^{\infty} \omega(\xi) f(x - \xi) d\xi, \tag{13.2}$$

will be called an *averaged* (smoothed) function. Here $\omega(\xi)$ is a weight function, which acts as a filter. The expression (13.2) means that the value of the averaged function $\bar{f}$ at point x depends on values of the original function f at all points of the domain of integration and each of these values is taken into account with its own weight ω.

It is easy to see that functions f and ω play equal roles in the expression (13.2), *i.e.*, the averaged function may also be written as

$$\bar{f}(x) = \int_{-\infty}^{\infty} f(\xi) \omega(x - \xi) d\xi. \tag{13.3}$$

Indeed, let $\eta = x - \xi$ then

$$d\xi = -d\eta, \quad f(\xi) = f(x - \eta), \quad \omega(x - \xi) = \omega(\eta).$$

Hence,

$$f(x) = -\int_{\infty}^{-\infty} f(x - \eta) \omega(\eta) d\eta = \int_{-\infty}^{\infty} \omega(\eta) f(x - \eta) d\eta,$$

that with the accuracy to notation coincides with the (13.3).

Meanwhile, the function ω which we consider as a smoothing function (filter) must meet some obvious requirements:

1. Since the average value of a constant must be equal the constant itself (*i.e.*, if f= const = C then f= $\bar{f}$ = C), we obtain

$$C = \int_{-\infty}^{\infty} \omega(\xi) C d\xi = C \int_{-\infty}^{\infty} \omega(\xi) d\xi,$$

or

$$\int_{-\infty}^{\infty} \omega(\xi) d\xi = 1.$$

2. The function $\overline{f}$ exists if the improper integrals in (13.2), (13.3) converge. And they converge, if for $|\xi| \to \infty$ the weight function $\omega(\xi)$ is such that $\omega(\xi) f(x\text{-}\xi) \to 0$ no slower than $1/\xi^{1+\alpha}$, $\alpha > 0$. Indeed, it is known that the integral $\int_{-\infty}^{\infty} \frac{1}{\xi} d\xi$ diverges, while the integral $\int_{-\infty}^{\infty} \frac{1}{\xi^{1+\alpha}} d\xi$ converges for any arbitrarily small positive α (see., *e.g.*, [2]).

In practice, the spatial or temporal averaging is used most commonly. As a filter $\omega(\xi)$ an even non-negative function given by the expression

$$\omega(\xi) = \begin{cases} \frac{1}{L}, & |\xi| < \frac{L}{2}, \\ 0, & |\xi| \geq \frac{L}{2}, \end{cases} \tag{13.4}$$

is taken. The domain $\left(-\frac{L}{2}, \frac{L}{2}\right)$ is the so-called *averaging interval*. This is either a time interval, or some spatial range. The average values of the function $f(x)$ are calculated, using the formula

$$\overline{f}(x) = \frac{1}{L} \int_{-\frac{L}{2}}^{\frac{L}{2}} f(x - \xi) d\xi = \frac{1}{L} \int_{x-\frac{L}{2}}^{x+\frac{L}{2}} f(\eta) d\eta.$$

Under such averaging, the Reynolds'conditions $1 - 4$) (see, p.159) are satisfied exactly, whereas the condition 5) only approximately: the function $\overline{f}$ depends on the averaging interval. For example, the condition 4) holds due to definition (13.2) or (13.3): the averaging procedure is linear, $\overline{f_1 + f_2} = \overline{f_1} + \overline{f_2}$. However, the condition 5a) fails to hold and the result of the second averaging differs from the first one (from the single averaging). Indeed

$$\overline{\overline{f}} = \int_{-\infty}^{\infty} \omega(\eta) \left(\int_{-\infty}^{\infty} \omega(\xi) f(x - \xi - \eta) d\xi \right) d\eta.$$

Designating $\zeta = \xi + \eta$ we obtain $\eta = \zeta - \xi$, $d\eta = d\zeta$ and further

$$\bar{\bar{f}} = \int_{-\infty}^{\infty} \omega(\eta) \int_{-\infty}^{\infty} \omega(\xi) f(x - \zeta) d\xi d\eta$$

$$= \iint_{-\infty}^{\infty} \omega(\zeta - \xi) \omega(\xi) f(x - \zeta) d\xi d\zeta$$

$$= \int_{-\infty}^{\infty} \left(\int_{-\infty}^{\infty} \omega(\zeta - \xi) \omega(\xi) d\xi \right) f(x - \zeta) d\zeta$$

$$= \int_{-\infty}^{\infty} \omega_1(\zeta) f(x - \zeta) d\zeta.$$

Thus, repeated smoothing is equivalent to the single smoothing, but with another weight function ω_1

$$\omega_1(\zeta) = \int_{-\infty}^{\infty} \omega(\zeta - \xi) \omega(\xi) d\xi.$$

When a rectangular weight function (13.4) is used, the second smoothing leads to averaging with the triangular weight function $\omega_1 = \omega^{(2)}$ (Fig. **13.1**)

$$\omega^{(2)}(\zeta) = \int_{-\infty}^{\infty} \omega(\xi) \omega(\zeta - \xi) d\xi$$

$$= \frac{1}{L} \int_{-\frac{L}{2}}^{\frac{L}{2}} \omega(\zeta - \xi) d\xi = \frac{1}{L^2}(L - |\zeta|), \quad |\zeta| < L,$$

$$\omega^{(2)}(\zeta) = 0, \quad |\zeta| \geq L.$$

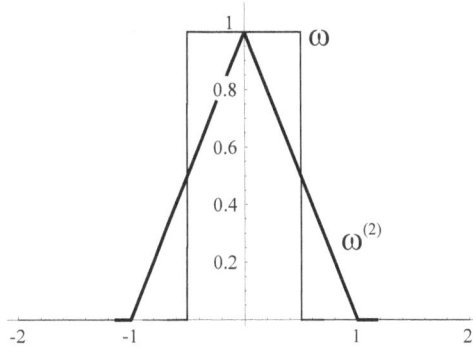

Fig. (13.1). Repeated smoothing with rectangular filter.

The finiteness of the averaging interval leads to dependence on the magnitude of L of the result of smoothing. This threatens the very definition of the average, as a kind of an objective characteristic of the flow, independent of the observer.

Therefore, choosing L, one should strive to minimize the dependence of $\overline{f}$ on the averaging interval. If this may be achieved then the meaningful definition of the mean value is possible. A detailed discussion of this important issue may be found, for example, in [8] or [9].

13.4. THE REYNOLDS EQUATION

In order to obtain equations describing evolution of the average characteristics of the flow, it is necessary to average the equations for the corresponding instantaneous values. We shall study the simplest case, when turbulence is generated exclusively by velocity fluctuations.

The averaged form of the continuity equation is easy to find. We represent velocity as a sum of averaged and fluctuating components $\vec{v} = (\vec{\overline{v}} + \vec{v}')$, substitute this into the equation (6.18) and average the result

$$\overline{\partial_t \rho + \left(\nabla, \rho(\vec{\overline{v}} + \vec{v}')\right)} = 0.$$

Next, using the properties of the averaging operator, we obtain

$$0 = \partial_t \rho + \left(\nabla, \rho(\overline{\vec{\overline{v}}} + \overline{\vec{v}'})\right) = \partial_t \rho + (\nabla, \rho\vec{\overline{v}}), \tag{13.5}$$

or in case of incompressible fluid

$$(\nabla, \vec{\overline{v}}) = 0. \tag{13.6}$$

Exercise. Show that the velocity fluctuations satisfy the continuity equation.

It is easy to see that the averaged equations (13.5) and (13.6) to accuracy of notation coincide with their non-averaged analogs (6.18) and (6.20). This is not surprising, since the continuity equation is linear with respect to the averaged quantity (velocity vector). The emergence of new terms to be expected when averaging nonlinear equations. The fluid model contains such equation: it is the equation of motion. Arising from it the kinetic energy balance equation also contains a nonlinear term.

Firstly, consider the case of compressible fluid and average the Navier-Stoke-
-Duhem equation in the form (11.8). Transforming the left-hand side with the help
of the continuity equation (6.18), we write it in the form[1]

$$\partial_t \rho \vec{v} + \nabla_k (\rho \vec{v} v_k) + \nabla p - 2\mu \, div \mathrm{D} - \rho \vec{b} = 0. \qquad (13.7)$$

Now we write down the velocity as the sum of the averaged and fluctuating
components $\vec{v} = (\bar{\vec{v}} + \vec{v}')$ and average the resulting equation on example of its
i-th components:

$$0 = \overline{\partial_t \rho (\bar{v}_i + v'_i) + \nabla_k (\rho (\bar{v}_i + v'_i)(\bar{v}_k + v'_k))} +$$
$$+\overline{\nabla_i p - \mu \nabla_k (\nabla_k (\bar{v}_i + v'_i) + \nabla_i (\bar{v}_k + v'_k)) - \rho b_i}. \qquad (13.8)$$

Again we use the properties of the averaging operator to transform this equation
to the form

$$\partial_t \rho \bar{v}_i + \nabla_k (\rho \bar{v}_i \bar{v}_k) = -\nabla p + 2\mu \nabla_k \bar{\mathrm{D}}_{ik} - \nabla_k (\overline{\rho v'_i v'_k}) - \rho b_i, \qquad (13.9)$$

where $\bar{\mathrm{D}}_{ik} \equiv \frac{1}{2}(\nabla_k \bar{v}_i + \nabla_i \bar{v}_k)$.

If the fluid is incompressible, the equation is simplified somewhat due to the fact
that

$$\mu \nabla_k \bar{\mathrm{D}}_{ik} = \mu \Delta \bar{v}_i + \mu \, \partial_{x_i} (\nabla, \bar{\vec{v}}) \overset{(13.6)}{=} \mu \Delta \bar{v}_i$$

As a result, we obtain the evolution equation of the averaged velocity vector of
the incompressible fluid or the *Reynolds equation*:

$$\partial_t \rho \bar{\vec{v}} + \nabla_k (\rho \bar{\vec{v}} \bar{v}_k) = -\nabla p + \mu \Delta \bar{\vec{v}} - \nabla_k (\overline{\rho \vec{v}' v'_k}) + \rho \vec{b}. \qquad (13.10)$$

Subtracting from the left side of (13.10) the continuity equation of the mean
motion (13.6), multiplied by $\bar{\vec{v}}$ (this is the inverse of the trick that we have just
used), we get another form of the Reynolds equation:

$$\rho \, \partial_t \vec{\bar{v}} + \rho \bar{v}_k \nabla_k \vec{\bar{v}} = -\nabla p + \mu \Delta \vec{\bar{v}} - \nabla_k (\overline{\rho \vec{v}' v'_k}) + \rho \vec{b}. \qquad (13.11)$$

A new term on the right-hand side of the equation (13.11) is the divergence of the tensor R whose components

$$R_{ij} = \overline{\rho v_i' v_j'}$$

are correlations of fluctuating components of the velocity vector. This tensor is called the *Reynolds stress tensor*. Thus, the exchange of momentum between fluid "particles" in the turbulent flow is carried out not only by the molecular viscosity (the term $\mu \Delta \vec{\bar{v}}$), but also due to mixing produced by the velocity fluctuations (the term $-\nabla_k (\overline{\rho \vec{v}' v'_k})$).

If we assume that turbulence is a result of transition of a part of energy of the averaged flow, to small-scale perturbations, the characteristics of turbulence should depend on the field of the averaged velocity. This suggests to describe the Reynolds stress tensor in terms of the gradient of the averaged velocity by analogy with description of the viscous stress tensor in terms of the velocity gradient. Suppose by definition

$$R_{ij} \equiv -2\mu_T \bar{D}_{ij} = -\mu_T \left(\partial_{x_i} \bar{v}_j + \partial_{x_j} \bar{v}_i \right), \qquad (13.12)$$

where the factor μ_T is called *eddy viscosity*. After substituting (13.12) in (13.11), and taking into account the continuity equation of the incompressible fluid, the Reynolds equation looks as follows

$$\rho \, \partial_t \vec{\bar{v}} + \rho \bar{v}_k \nabla_k \vec{\bar{v}} = -\nabla p + (\mu + \mu_T) \Delta \vec{\bar{v}} + \rho \vec{b} \qquad (13.13)$$

or in a more general form which will be useful further,

$$\rho \, \partial_t \vec{\bar{v}} + \rho (\vec{\bar{v}}, \nabla) \vec{\bar{v}} = -\nabla p + 2(\mu + \mu_T) div \bar{D} + \rho \vec{b}. \qquad (13.14)$$

Since, $\mu_T \gg \mu$, the factor $(\mu + \mu_T)$ in the equation (13.13), is usually replaced by the eddy viscosity.

The factor μ describes physical properties of the fluid (the statistical properties of the random molecular motion), and the factor μ_T describes the statistical properties of fluctuations. It is easy to see that the current approach to describe turbulence actually is the introduction of the second amendment R to the stress tensor T

$$T = -pI + T' + R = -pI + 2\mu\overline{D} + 2\mu_T\overline{D}.$$

However, the tensor $\overline{D}$ here is the symmetric part of the gradient of the averaged velocity $\nabla\overline{v}$, rather than the instantaneous one, as in the Navier-Stokes equation.

13.5. THE ENERGY BALANCE

Along with mass, the energy is another scalar characteristic of a body. If the mass characterizes the ability of a body to resist accelerating, the energy of a moving body characterizes its ability to do work. These two quantities would be sufficient if the fluid motion of arbitrary scales would be smooth and orderly. However, now we are studying another case, and the observed chaotic motion forces us to abandon the detailed description of motion of all scales and limit it to the averaged characteristics.

Besides, we cannot simply neglect the chaotic components of the motion (it would have cost us the conservation laws, *i.e.* the bases of the theory), but we are trying to take them into account parametrically. We did that, with random fluctuations at the molecular level, by introduction of the internal energy. We have considered the total kinetic energy of the motion of all scales, picked out the part, which is related to microscopic and, therefore, the most high-frequency movements, called it the internal energy and used it for approximate description of all movements at this level. However, if the density of kinetic energy at a certain point in the space-time is always associated with the velocity vector at that point, then the density of internal energy is not connected with any specific velocity vector. This numerical function describes the average ability of microscopic movements to do work.

Now, when the loss of stability of the laminar flow was discovered and errors in description of kinematics presented in sec.5.2 became inadmissible, to move forward, we again forced to abandon the explicit descriptions of movements of a part of scales. We allocate a share of the total energy of the motion, call it the *turbulent energy*, and associate it with all fluctuations which we are unable to describe explicitly. Thus, previously general energy of the motion is now broken up into three parts: the *kinetic energy of mean motion* K_A, the *turbulent energy* K_T and the *internal energy E*.

All these parts correspond to different scales of motion and all together form what is, as before, the *total energy*. It is still a conserved quantity, because it is a measure of the ability of the body to perform work irrespective of the scale of the motion from the very large, up to the mechanical interactions of molecules. The averaging of the total energy balance equation (11.20) gives a similar equation for turbulent media

$$d_t((K_A + K_T) + U + E) = \int_\chi div(\overline{T\vec{v}})dV = \int_{\partial\chi} \left(\overline{T\vec{v}}, \vec{n}\right) dS. \qquad \textbf{(13.15)}$$

Since, the right-hand side of (13.15) describes the flux through the surface of the world-tube of the body, which is equal to zero by definition, the total energy of the body does not change. Components of the total energy, on the contrary, are conserved only in specific cases.

For each type of the energy density it is possible to write down its own differential balance equation, and in sum they have to give the total energy conservation law. To obtain these equations, we proceed as follows:

1. Calculate the scalar product of the Reynolds equation in the form (13.14) and the average velocity vector

$$\left(\overline{v}, \rho\, \partial_t \overline{v} + \rho(\overline{v}, \nabla)\overline{v} + \nabla p - 2(\mu + \mu_T)div\overline{D} - \rho\vec{b}\right) = 0$$

and obtain the kinetic energy density balance equation of mean motion $k_A \equiv \frac{1}{2}\rho(\overline{v}, \overline{v})$:

$$\rho\, \partial_t \frac{k_A}{\rho} + (\rho\overline{v}, \nabla\frac{k_A}{\rho}) = -(\overline{v}, \nabla p) + 2(\mu + \mu_T)(\overline{v}, div\overline{D}) + \rho(\overline{v}, \vec{b}). \qquad \textbf{(13.16)}$$

Next, we transform it such that the right side contains only dissipative terms. Firstly, we consider the incompressible medium, and write down

$$(\overline{v}, \nabla p) = (\nabla, p\overline{v}).$$

The body force density is expressed through the potential Ф. The term $\rho(\overline{v}, \vec{b})$ is transformed in the same way as on p.100, and using the continuity equation we obtain

$$\rho(\vec{v}, \vec{b}) = -\rho(\vec{v}, \nabla\Phi) - \rho\,\partial_t\Phi - \Phi\underbrace{(\partial_t\rho + (\nabla, \rho\vec{v}))}_{=0} =$$

$$= -\partial_t(\rho\Phi) - (\nabla, \rho\Phi\vec{v}) = -d_t(\rho\Phi),$$

Finally, the scalar product in the second term on the right-hand side (13.16) is transformed according to the formula

$$(\overline{\vec{v}}, div\overline{D}) = (\nabla, \overline{D}\overline{\vec{v}}) - \overline{D}:\overline{D}. \tag{13.17}$$

Substituting all this in the original equation (13.16), we have

$$\partial_t(k_A + \rho\Phi) + \left(\nabla, (k_A + \rho\Phi + p)\overline{\vec{v}} - 2(\mu + \mu_T)\overline{D}\overline{\vec{v}}\right) = \tag{13.18}$$
$$= -2(\mu + \mu_T)\overline{D}:\overline{D}.$$

2. Now take the equation (11.23), which describes the balance of the kinetic energy density of the non-averaged motion $k \equiv \frac{1}{2}\rho(\vec{v}, \vec{v})$. We write it as

$$\partial_t(k + \rho\Phi) + (\nabla, (k + p + \rho\Phi)\vec{v} - 2\mu D\vec{v}) = -2\mu D : D,$$

substitute $\vec{v} = (\overline{\vec{v}} + \vec{v}')$, average and obtain

$$\overline{\partial_t\left(\frac{1}{2}\rho(\overline{v}_i + v_i')^2 + \rho\Phi\right)} +$$
$$+ \overline{\left(\nabla, \left(\frac{1}{2}\rho(\overline{v}_i + v_i')^2 + p + \rho\Phi\right)(\overline{\vec{v}} + \vec{v}') - 2\mu(\overline{D} + D')(\overline{\vec{v}} + \vec{v}')\right)} =$$
$$= -2\mu\overline{(\overline{D} + D'):(\overline{D} + D')},$$

where $D'_{ik} \equiv \frac{1}{2}(\nabla_k v'_i + \nabla_i v'_k)$. Taking into account the properties of the averaging operator, we find

$$\partial_t(k_A + \rho\Phi + k_T) +$$
$$+ \nabla_k\left((k_A + \rho\Phi + p + k_T)\overline{v}_k + \rho\overline{v_i' v_i' v_k'}\right) - \tag{13.19}$$
$$- \nabla_k\left(2\mu(\overline{D}_{ki}\overline{v}_i + \overline{D'_{ki}v'_i})\right) = -2\mu(\overline{D}:\overline{D} + \overline{D':D'}).$$

The quantity denoted by $k_T \equiv \frac{1}{2}\rho \overline{v_i' v_i'}$ is called the *turbulent energy density*. We have derived the equation describing the balance of the sum of kinetic and turbulent energy densities. Subtracting the kinetic energy density balance equation of mean motion (13.18) (corresponding terms are underlined), we find the turbulent energy density balance equation:

$$\partial_t k_T + \left(\nabla, k_T \overline{\vec{v}} + \frac{1}{2}\rho \overline{v_i' v_i' \vec{v}'} + 2\mu_T \overline{D}\overline{\vec{v}} - 2\mu \overline{D'\vec{v}'} \right) = \tag{13.20}$$
$$= 2\mu_T \overline{D} : \overline{D} - 2\mu \overline{D' : D'}.$$

3. Consider, finally, the internal energy balance equation (11.26), together with the formula (11.18). As in the previous cases, we substitute the velocity for the sum of the mean and fluctuating components, and average. In view of the incompressible medium ($\nabla, \vec{v}$) = 0, we obtain:

$$\overline{\partial_t \varepsilon + (\nabla, \varepsilon(\vec{v} + \vec{v}')) - 2\mu(\overline{D} + D') : (\overline{D} + D')} = 0, \tag{13.21}$$

and, as a result, we find

$$\partial_t \varepsilon + (\nabla, \varepsilon \overline{\vec{v}}) = 2\mu \overline{D} : \overline{D} + 2\mu \overline{D' : D'}.$$

4. To make viewing the results easier, we write down together all equations obtained. Here they are

$$\partial_t (k_A + \rho\Phi) + (\nabla, \underbrace{(k_A + \rho\Phi + p)\overline{\vec{v}}}_{(1)} - \underbrace{2\mu_T \overline{D}\overline{\vec{v}}}_{(2)} - \underbrace{2\mu \overline{D}\overline{\vec{v}}}_{(3)})$$
$$= -2(\mu + \mu_T)\overline{D} : \overline{D},$$
$$\partial_t k_T + (\nabla, \underbrace{k_T \overline{\vec{v}}}_{(1)} + \underbrace{\frac{1}{2}\rho \overline{v_i' v_i' \vec{v}'} + 2\mu_T \overline{D}\overline{\vec{v}}}_{(2)} - \underbrace{2\mu \overline{D'\vec{v}'}}_{(3)})$$
$$= 2\mu_T \overline{D} : \overline{D} - 2\mu \overline{D' : D'},$$
$$\partial_t \varepsilon + (\nabla, \underbrace{\varepsilon \overline{\vec{v}}}_{(1)}) = 2\mu \overline{D} : \overline{D} + 2\mu \overline{D' : D'}.$$

The left-hand sides of these equations contain divergences of the flux densities of corresponding components of the energy. They include the transfer by the averaged flow (1), by turbulent fluctuations (2) and by viscous forces (3). The right-hand sides are sources or sinks of the components of the total energy k_A, k_T and ε, and since the total energy is conserved, the right sides, in fact, describe the exchange between these components. Thus, we see that the internal energy is powered due to dissipation of both the kinetic energy and the turbulent energy. In turn, the turbulent energy is fed by a part of the kinetic energy of mean motion. In sum, all three equations give the total energy density balance equation

$$\partial_t(k_A + k_T + \varepsilon + \rho\Phi) + \left(\nabla, (k_A + k_T + \varepsilon + \rho\Phi + p)\vec{\bar{v}}\right) + $$
$$+ \left(\nabla, \frac{1}{2}\rho\overline{v_i'v_i'\vec{v}'} - 2\mu(\overline{\mathrm{D}\vec{\bar{v}}} + \overline{\mathrm{D}'\vec{v}'})\right) = 0. \tag{13.22}$$

If we integrate it over a fixed volume V, then we obtain the integral balance equation of the total energy $((K_A + K_T) + U + E)$. To demonstrate this, we rewrite equation (13.22) in the equivalent form

$$\rho d_t \left(\frac{k_A + k_T + \varepsilon}{\rho} + \Phi\right) = -\left(\nabla, p\vec{\bar{v}} + \frac{1}{2}\rho\overline{v_i'v_i'\vec{v}'} - 2\mu(\overline{\mathrm{D}\vec{\bar{v}}} + \overline{\mathrm{D}'\vec{v}'})\right).$$

Integrating with respect to V and applying the Gauss' theorem to the right side, we find

$$d_t((K_A + K_T) + U + E) = -\int_{\partial V} (p\vec{\bar{v}} + \frac{1}{2}\rho\overline{v_i'v_i'\vec{v}'} - 2\mu(\overline{\mathrm{D}\vec{\bar{v}}} + \overline{\mathrm{D}'\vec{v}'}))_n dS.$$

The right side is the integral flux of the total energy through the boundary of the volume V. If it is zero (*i.e.*, the influx of the total energy is equal to its outflux), the total energy in the volume is conserved. Otherwise, the difference is balanced by corresponding change of the total energy.

NOTES

[1]This transformation is described in the footnote on p.142.

Classical Fluid Mechanics, 2017, 171-185

CHAPTER 14

Boundary Layers

Abstract: The important concept of the boundary layer is introduced. We study the laminar, thermal and turbulent boundary layers and also derive a system of boundary layer equations. The concept of the boundary layer thickness is considered and the separation of the boundary layer is discussed. The general profile of the average velocity of the turbulent flow near a smooth wall, and the effects of roughness are being studied. The notion of the roughness parameter is defined.

Keywords: Boundary layer, Boundary layer separation, Boundary layer thickness, Friction velocity, Prandtl number, Roughness parameter, Thermal boundary layer, Turbulent boundary layer, Viscous sublayer.

14.1. TH LAMINAR BOUNDARY LAYER

Consider a simple problem: a flow of a fluid along a flat solid surface. If the Reynolds number Re is large (either the scale L is large (far from the wall), and/or the velocity scale U is high), the fluid motion may be described by the perfect fluid model. On the other hand, it is clear that the wall retards fluid flow and, if we take the no-slip boundary condition, the flow velocity at the surface should be zero. Thus, near the wall the number Re is small and the model of the perfect fluid stops working and the viscous fluid model should be used instead. However, far from the wall the flow is determined by inertial forces and viscous forces are negligible. Taking them into account here is a waste of effort. What to do?

In accordance with the Prandtl's[1] suggestion, we select a thin layer of fluid, adjacent to the solid boundary, within which the viscous forces dominate or are comparable with the inertial forces. This area is called the *boundary layer*. The fluid velocity here rapidly increases from zero on the solid surface up to values on the external boundary of the layer, which are determined by the fluid flow outside the boundary layer. Such an approach allows us to kill two birds with one stone: firstly, we may use the Euler equations in a region with large values of Re and, secondly, we take viscosity into account only where it is significant, *i.e.*, in a thin layer near the solid surface, which allows simplification of equations of the

Michael Belevich

viscous fluid model. The concept of the boundary layer proved to be very fruitful and is widely used. So, boundary layers are often considered not only on solid surfaces, but also at interfaces between two media. For example, on both sides of the air-water interface. In all cases, this area is characterized by large velocity gradients, and is an area of intense vortex motion.

14.1.1. The Thickness of the Boundary Layer

Consider the simplest case of the boundary layer formation on a semi-infinite flat thin plate in a uniform flow moving with velocity U. Let's choose the Cartesian coordinates (x, y, z) such that the plate is described by expressions $x \geq 0, z = 0$. The x-axis is oriented in the direction of the incoming flow velocity. Besides, movements along the y-axis are absent, and the problem may be regarded as two-dimensional with coordinates (x, z). The incoming flow is potential, *i.e.*, irrotational.

The model of the perfect fluid is insensitive to the introduction of a plate into the flow. In order its presence could affect the fluid motion, it is necessary to allocate an area along the plate, where other means of description of a moving medium are used. This area is the boundary layer, within which the fluid flow is braked sharply. Since the problem is symmetric with respect to the x-axis we shall consider the upper half-plane only.

The thickness of the boundary layer δ is not constant and increases from zero with the distance from the edge of the plate. To estimate the value of δ consider parameters which determine this quantity. They are only three: t, v, U and they have following physical dimensions:

$$[t] = s, \quad [v] = \frac{m^2}{s}, \quad [U] = \frac{m}{s}.$$

The velocity of incoming flow U is the velocity scale for horizontal movements. The ratio of the other two parameters has dimension of the square of velocity $[v/t] = [U]^2$ and is related to the kinematic viscosity v. Obviously, the square root of this quantity may be considered as the scale for the vertical velocity $w \equiv \sqrt{v/t}$. Indeed, if viscosity is absent $v = 0$, the flow is insensitive to the presence of walls and $w = 0$. If on the contrary $v \to \infty$, the plate slows the entire flow and $w \to \infty$. The decelerating effect of the plate, as we see, extends up the faster, the higher the value of viscosity.

Perturbations, which are introduced into the flow by the plate, are spreading up with characteristic velocity w and are drifting with the flow with characteristic velocity U. During time t perturbations from the plate reach the height $\delta_y = wt$. During the same time, they are drifting with the flow at the distance $x = Ut$. Hence, for the boundary layer thickness we find

$$\delta_y = wt = \sqrt{\nu t} = \sqrt{\nu \frac{x}{U}} = x \sqrt{\frac{\nu}{Ux}} = \frac{x}{\sqrt{Re}}, \tag{14.1}$$

i.e. with the distance x from the edge of the plate the boundary layer thickness increases as $\sqrt{x}$.

If the flow velocity U is large, then Re is large, δ_y is small and vortex disturbances generated by the plate, barely penetrate deep into the flow. In this case, we may assume that the effect of the boundary layer is absent and the flow remains potential.

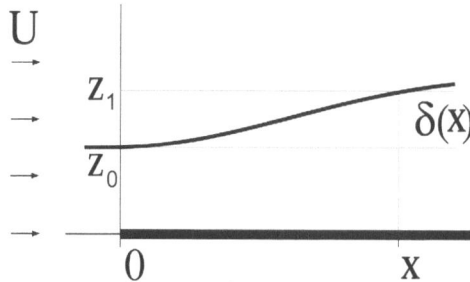

Fig. (14.1). Displacement thickness.

However, it should be kept in mind that a boundary layer does not have a distinct border, and therefore its thickness is a relative concept. There are different viewpoints on what should be called the border of a boundary layer. For example, a thickness of the boundary layer δ_y may be understood as a distance from a solid surface along the normal, where u-component of velocity reaches a certain share of the flow velocity, say, 99%.

Another quantity characterizing the transverse scale of the boundary layer, is the so-called displacement thickness δ_* defined by the formula

$$\delta_* = \int_0^\infty \left(1 - \frac{u(x,z)}{U}\right) dz. \tag{14.2}$$

The meaning of this definition is that due to friction on the plate, the longitudinal component of velocity within the boundary layer decreases, and this, in turn, causes distortion of streamlines, which manifests itself as the ousting out of the incoming flow (Fig. **14.1**). Since the fluid flow through any section of the stream tube must be the same (see, p. 131), and u-component of the velocity decreases, the stronger the greater the distance from the surface of the plate, the section of the stream tube should increase with increasing x.

In front of the plate the flow through a contour between streamlines passing through points $(0, 0)$ and $(0, z_0)$, is equal to

$$\int_0^{z_0} U dz = U z_0.$$

At a distance x from the edge of the plate, the fluid flow between the same streamlines is the same

$$\int_0^{z_1(x)} u(x, z) dz = \int_0^{z_0} U dz. \tag{14.3}$$

Here $z = z_1(x)$ is the equation of the streamline passing through the point $(0, z_0)$. The latter integral in (14.3) may be written as

$$\int_0^{z_0} U dz = \int_0^{z_1(x)} U dz - \int_{z_0}^{z_1(x)} U dz = \int_0^{z_1(x)} U dz - (z_1(x) - z_0)U. \tag{14.4}$$

Substituting (14.4) in (14.3), we find

$$\delta(x) = (z_1(x) - z_0) = \int_0^{z_1(x)} \left(1 - \frac{u(x,z)}{U}\right) dz = \int_0^{z_0 + \delta(x)} \left(1 - \frac{u(x,z)}{U}\right) dz.$$

When $z_0 \to \infty$, we obtain (14.2).

The displacement leads to the fact that the pattern of streamlines becomes distorted also outside the boundary layer. Thus, the boundary layer has an effect on the external flow. In particular, due to displacement, the vertical component of velocity does not tend to zero when $z \to \infty$.

14.1.2. Boundary Layer Equations

Plane-parallel flow around an infinitely wide plate of the length L, as it was noted, is two-dimensional. Non-dimensional equations of the viscous fluid in this case, are as follows:

$$\partial_x u + \partial_z w = 0, \tag{14.5}$$

$$\partial_t u + u\,\partial_x u + w\,\partial_z u = -\frac{1}{\rho}\partial_x p + \frac{1}{Re}(\partial_{xx}u + \partial_{zz}u), \tag{14.6}$$

$$d_t v = 0, \tag{14.7}$$

$$\partial_t w + u\,\partial_x w + w\,\partial_z w = -\frac{1}{\rho}\partial_z p + \frac{1}{Re}(\partial_{xx}w + \partial_{zz}w). \tag{14.8}$$

Direction of x-axis still coincides with the velocity vector U of the incoming flow, and z-axis is directed vertically upward. We are interested in the motion of the fluid near the plate in a thin layer with thickness $\sim \frac{1}{\sqrt{Re}}$. Using the coordinate change we shall pass to consideration of motion in a region independent on the characteristic of the flow Re. Let $\zeta = z\sqrt{Re}$ Then

$$w = d_t z = \frac{1}{\sqrt{Re}} d_t \zeta = \frac{1}{\sqrt{Re}} W, \qquad W \equiv d_t \zeta,$$

and equations (14.5)–(14.8) may be written as:

$$\partial_x u + \partial_\zeta W = 0,$$

$$\partial_t u + u\,\partial_x u + W\,\partial_\zeta u = -\frac{1}{\rho}\partial_x p + \frac{1}{Re}(\partial_{xx}u + Re\,\partial_{\zeta\zeta}u),$$

$$\frac{1}{\sqrt{Re}}\partial_t W + \frac{u}{\sqrt{Re}}\partial_x W + \frac{W}{\sqrt{Re}}\partial_\zeta W = -\frac{\sqrt{Re}}{\rho}\partial_\zeta p + \frac{1}{Re^{3/2}}\partial_{xx}W + \frac{1}{\sqrt{Re}}\partial_{\zeta\zeta}W.$$

Normalizing the latter equation by $\sqrt{Re}$, we rewrite the last two equations as follows:

$$\partial_t u + u\,\partial_x u + W\,\partial_\zeta u = -\frac{1}{\rho}\partial_x p + \frac{1}{Re}\partial_{xx}u + \partial_{\zeta\zeta}u, \tag{14.9}$$

$$\frac{1}{Re}\left(\partial_t W + u\,\partial_x W + W\,\partial_\zeta W\right) = -\frac{1}{\rho}\partial_\zeta p + \frac{1}{Re^2}\partial_{xx}W + \frac{1}{Re}\partial_{\zeta\zeta}W. \tag{14.10}$$

In the limit $Re \to \infty$ we obtain the system of boundary layer equations:

$$\partial_x u + \partial_\zeta W = 0, \tag{14.11}$$
$$\partial_t u + u\,\partial_x u + W\,\partial_\zeta u = -\frac{1}{\rho}\partial_x p + \partial_{\zeta\zeta}u,$$

$$0 = -\frac{1}{\rho}\partial_\zeta p. \tag{14.12}$$

According to (14.12) the pressure is constant over the cross-section of the boundary layer and, therefore, must be equal to the pressure in the incoming flow which is described *via* the model of the perfect fluid. Thus, the equality

$$\partial_t U + U\,\partial_x U = -\frac{1}{\rho}\partial_x p \tag{14.13}$$

holds and the horizontal component of pressure gradient is a known function. Returning to original coordinates, we obtain

$$\partial_x u + \partial_z w = 0, \tag{14.14}$$
$$\partial_t u + u\,\partial_x u + w\,\partial_z u = -\frac{1}{\rho}\partial_x p + \frac{1}{Re}\partial_{zz}u.$$

This system of equations with boundary conditions

$$u|_{z=0} = w|_{z=0} = 0, \qquad u|_{z\to\infty} \to U$$

and corresponding initial condition allows one to find components of the velocity vector in the domain $x \in [0, L]$, $z \in [0, \infty]$.

14.1.3. Separation of the Boundary Layer

Boundary layer equations were derived for the case of a flow over an infinite flat plate. It may be shown, however, that in the first approximation, they are fair also in the description of the flow over two-dimensional curved surfaces with small curvature. The radius of curvature should be large, everywhere compared to the boundary layer thickness (see, [10]). Meanwhile the curvature affects the velocity profile, modifies it, and may give rise to the phenomenon of *boundary layer separation*.

Consider a stationary flow over a circular cylinder and perpendicular to its axis. Let radius of a cross-section of the cylinder and velocity of the incoming flow are large enough so that the boundary layer equations are applicable. On the surface of the cylinder a boundary layer is formed and outside this layer the fluid may be considered as the perfect fluid whose motion is potential.

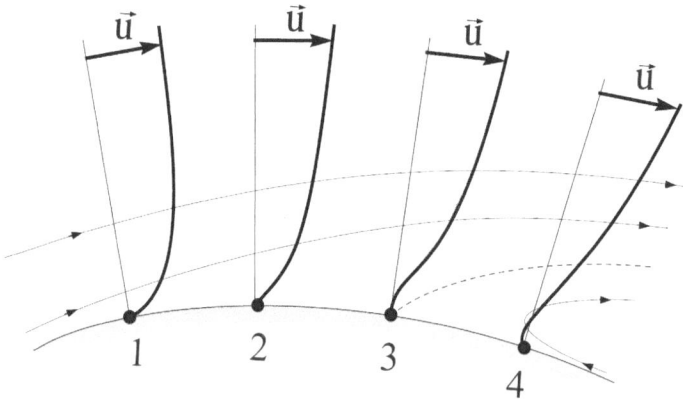

Fig. (14.2). Separation of the boundary layer.

Horizontal component of the pressure gradient in the stationary conditions is equal to

$$-\frac{1}{\rho}\partial_x p = u\,\partial_x u, \tag{14.15}$$

(cf. Eq. (14.13)) where u is the horizontal component of the velocity vector of the incoming flow (see. Fig. **14.2**). Above the upper point (2) of the cross-section of the cylinder the value of u reaches its maximum $|u| = |U|$ and $\partial_x u = 0$, and hence, $\partial_x p = 0$. Thus, the value of u increases from point (1) to point (2) and then falls. The pressure, on the contrary, decreases to a point (2), and then increases.

Inasmuch as the pressure gradient $\partial_x p$ on the plot (2) – (4) is positive, the velocity gradient $\partial_x u$ here is negative, fluid slows down and at the point (3) the velocity drops to zero $u = 0$. Starting from the point (3), the fluid flows in the opposite direction. At point (3) the boundary layer separates from the surface, pushed aside by the return flow. The interface (shown by a dotted line) is unstable and quickly rolls up into a vortex. The second such vortex is formed in the lower half of the cylinder. They alternately are detached from the surface of the cylinder, are carried away by the flow, and in their place new vortices are formed. A trail behind an obstacle, which consists of successive oppositely swirling vortices, is called the Kármán vortex street[2]. This beautiful phenomenon often may be seen in nature.

The necessary condition for the separation of the boundary layer, as we have seen, is the existence of sites on the surface of the body where the pressure gradient changes its sign. This condition is also satisfied in some other cases. For example, the boundary layer separation may occur in a flow inside an expanding tube.

14.2. THERMAL BOUNDARY LAYER

As has been said, if a flow is characterized by large values of Pe the evolution of the temperature field is provided mainly by the velocity field $\vec{v}$. However in those places in the flow over solid surfaces, where the number Re dramatically decreases due to reducing of the flow velocity, values of the number Pe should also decrease sharply. Terms describing the molecular heat conduction has here the same order of magnitude as terms describing the convective heat transfer and, thus, the influence of molecular heat conduction cannot be neglected. The problem, similar to the accounting of viscosity near interface, occurs and it is solved similarly. We select a border region of sharp changes in temperature and call it the *thermal boundary layer*. Within this layer the effect of molecular heat conduction is taken into account. However, a small thickness of this layer allows significant simplification of equations (similar to equations of the boundary layer).

Arguing, as in case of the viscous boundary layer, it is possible to obtain an estimate of the thickness of the thermal boundary layer δ_T:

$$\delta_T = \frac{L}{\sqrt{Pe}}. \tag{14.16}$$

Comparing the thickness of the viscous boundary layer δ_v with the value δ_T we see that their ratio

$$\frac{\delta_v}{\delta_T} = \sqrt{\frac{Pe}{Re}} = \sqrt{\frac{v}{\kappa}} \tag{14.17}$$

depends solely on properties of the fluid (*i.e.* κ and v) and is not connected with the scales of its motion (L and U). The non-dimensional ratio $Pr \equiv \frac{v}{\kappa}$ is called the *Prandtl number*. For gases, this quantity is of order unity (0.72 in the air, $\frac{2}{3}$ in monatomic and $\frac{3}{4}$ in diatomic gases). For the majority of dropping fluids it is more than 1, and for liquid metals it is much less than 1.

The equation describing evolution of temperature in the temperature boundary layer may be obtained in the same manner as the equations of the boundary layer. Here is this equation

$$\partial_t T + u\,\partial_x T + w\,\partial_z T = \kappa\,\partial_{zz}T. \tag{14.18}$$

We have used all the same notation, coordinates and bases as in the derivation of the boundary layer equations.

Exercise. Derive the equation (14.18). Please note that this equation is written in the dimensional form .

14.3. THE TURBULENT BOUNDARY LAYER

14.3.1. Introduction

Extraordinary entanglement of trajectories and the concept of turbulent flow, as a motion with significant vorticity, suggests to regard it as a superposition of eddies of different sizes, intensity and orientation. The presence of shear of an average velocity leads to stretching of vortices and to strengthening of vorticity due to the energy of the main fluid flow. The largest eddies are of the scale of the main flow. Such scale may be equal to the diameter of a pipe, a jet, to the trail behind an obstacle, to the thickness of a boundary layer, and so on. These vortices are the basic energy consumers of the main flow and, in turn, they deliver energy to smaller vortices. And so, down to the scales at which energy is dissipated due to viscosity, or in other words, is converted to heat.

Turbulent flows fed by the energy of the main flow may be divided into two groups: free turbulent flows and near-wall flows. The first group includes the turbulent jets and trails, *i.e.*, turbulent regions, with no solid boundaries. The second group includes turbulent flow in pipes, channels, and in boundary layers. Here we consider properties of the flow in a turbulent boundary layer near the wall, which is parallel to the average flow velocity.

The structure of near-wall turbulence is directly determined by the proximity of a solid boundary. The turbulent boundary layer is a complex entity, where three areas are usually distinguished:

1. *the viscous sublayer,* adjacent to the wall, where the number Re rapidly decreases and falls to zero at the wall. The share of Reynolds stresses in the shear stress reduces, and it becomes viscous just near the wall;

2. the *transition region*;

3. the anisotropic, *fully turbulent flow* where sizes of vortices are comparable with the distance to the wall.

14.3.2. General View of a Mean Velocity Profile

Consider the case of a stationary plane-parallel flow with constant viscosity in the region $z > 0$, without pressure gradient. Under these assumptions, the first Reynolds equation has the form

$$\nu\,\partial_{zz}\overline{u} - \partial_z\overline{u'w'} = 0, \tag{14.19}$$

which implies that the shear stress is constant, does not depend on the distance to the wall

$$\tau(z) = \rho\nu\,\partial_z\overline{u} - \rho\overline{u'w'} = \tau_0 = const \tag{14.20}$$

and is equal to τ_0, which is the shear stress on the wall $z = 0$. Since the quantity $\overline{u'w'}$ is unknown, it is impossible to uniquely determine the velocity $\overline{u}(z)$ from the equation (14.19). However, we may apply the following reasoning, which is often called the *dimensional considerations*.

The averaged velocity $\overline{u}(z)$ at a distance z from the wall is determined only by the quantities τ_0, z, ρ and ν (see. (14.20)), *i.e.*,

$$\overline{u} = \overline{u}(\tau_0, z, \rho, \nu). \tag{14.21}$$

The number of these parameters may be reduced, given that the dimension of mass is included only in the dimension of τ_0 and ρ. The dimensions of $\overline{u}$, z and ν do not contain it. Hence, these parameters may appear in (14.21) only as the ratio $\frac{\tau_0}{\rho}$ with dimension of square velocity

$$\overline{u} = \overline{u}(z, \nu, \frac{\tau_0}{\rho}). \tag{14.22}$$

The quantity $v_* \equiv \sqrt{\frac{\tau_0}{\rho}}$ with dimension of velocity is called the *friction velocity* and is a natural scale for velocity in the near-wall flow. Therefore, we may write

$$\overline{u} = v_* f(z, \nu, v_*). \tag{14.23}$$

Since, the ratio $\frac{\overline{u}}{v_*}$ is non-dimensional, the function f in (14.23) should be non-dimensional function of its arguments also. Note that the quantities z, ν and v_* may form only one non-dimensional combination $\frac{z v_*}{\nu}$. Thus, the general form of the velocity profile should be written as follows

$$\frac{\overline{u}}{v_*} = f(\zeta), \qquad \zeta \equiv \frac{z v_*}{\nu}, \tag{14.24}$$

where $f(\zeta)$ is a universal function. The dependence (14.24) itself is called the *Prandtl's universal law of near-wall turbulence*.

The Prandtl's law is derived assuming that the wall is smooth. How a real wall should look like, for it may be considered smooth? From (14.24) it follows that the ratio $z_* = \frac{\nu}{v_*}$ is the length scale in the turbulent boundary layer. Evidently, if the average roughness height h at the wall is less than z_*, the wall may be considered smooth.

These considerations are verified by observations (see, *e.g.*, [8]), according to which, for $h \leq 4z_*$ the velocity profile does not depend on h. Roughness elements are fully immersed in the viscous sublayer (*i.e.*, the fluid layer, where viscous forces dominate), and in this case the wall is called *hydraulically smooth*. If $h > 60z_*$ the viscous sublayer is destroyed and the flow straight near the wall consists of vortices arising above irregularities. The molecular viscosity no longer influences the flow, and such wall is called a *completely rough* wall. The case $h \in (4z_*, 60z_*)$ is intermediate transitional region.

14.3.3. Flow Near a Smooth Wall

Consider again the case of hydraulically smooth wall, where the Prandtl's law (14.24) holds. The universal function f may be defined in two limit cases: for small and large values of ζ.

1. Small values of ζ. Since $\overline{u'w'} = 0$ at the surface of the wall, it is clear that for small ζ the flow is determined by the viscous stress whereas the Reynolds stress may be neglected. Then from (14.20) and the friction velocity definition (see, p.181) we have

$$\partial_z \overline{u} = \frac{v_*^2}{\nu} = const. \tag{14.25}$$

Integrating (14.25) with the assumption $\overline{u}|_{z=0} = 0$, we find

$$\overline{u}(z) = \frac{v_*^2 z}{\nu} \quad \Rightarrow \quad \overline{u} = v_* \zeta. \tag{14.26}$$

A fluid layer, where $\nu|\partial_z \overline{u}| \gg |\overline{u'w'}|$, is called the *viscous sublayer*. Here the velocity $\overline{u}$ increases linearly with the distance ζ from the wall.

2. Large values of ζ. Away from the solid walls the turbulent stress is much greater than the viscous stress and for large values of z we may assume that

$$\tau_0 = -\rho \overline{u'w'}. \tag{14.27}$$

Since the influence of viscosity is neglected, the vertical component of velocity gradient should not depend on ν, and is determined only by two remaining parameters v_* and z. Only one combination with the dimension of velocity gradient may be constructed of these parameters $\frac{v_*}{z}$ and the velocity gradient itself should be proportional to this ratio $\frac{v_*}{z}$:

$$\partial_z \overline{u} \sim \frac{v_*}{z}, \tag{14.28}$$

for sufficiently large values of z. The proportionality factor is traditionally written as $\frac{1}{\kappa}$, where κ is the so-called von Kármán's constant. Integrating (14.28), we obtain

$$\overline{u} = \frac{v_*}{\kappa} \ln \frac{\zeta}{\zeta_0}. \tag{14.29}$$

Numerical values of parameters κ and ζ_0 should be determined from experiments. According to observations, for $\zeta > 30$ the distribution of the average velocity with high accuracy is described by (14.29). In this case, the best agreement with experimental data was achieved for $\kappa = 0.4$ and $\zeta_0 = 0.11$ (see, *e.g.*, [8] for details). The fluid layer, within which the formula (14.29) holds, is called the *logarithmic boundary layer*. It is formed in flows in pipes, channels, above plates. In all these cases the expression (14.29) holds (see. Fig. **14.3**).

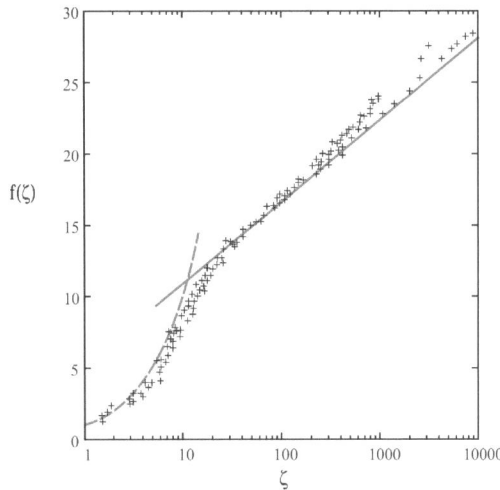

Fig. (14.3). The structure of the boundary layer [8]: the dependence of the universal function $f(\zeta)$ on the non-dimensional parameter ζ. "Pluses" mark the experimental data, solid line is the theoretical curve (14.29), and dotted line is the dependence (14.26).

Thus, when $\zeta < 5.$ the dependence (14.26) is valid. The range $\zeta \in (5., 30.)$ is the intermediate zone where terms $\rho \nu \, \partial_z \overline{u}$ and $-\rho \overline{u'w'}$ have the same order of magnitude. If we exclude this area out of our consideration (which probably will not lead to large errors), we may assume that the logarithmic boundary layer follows after the viscous sublayer. In this case, the viscous sublayer, where the velocity profile is described by the formula (14.26), extends to the value $\zeta = 11.1$, above which the logarithmic boundary layer is located and the dependence (14.29) holds.

In the previous chapter we have introduced the concept of turbulent viscosity and have associated components of the Reynolds' stress tensor with components of the gradient of average velocity (see Eq. (13.12)). In accordance with this formula we have

$$\rho \overline{u'w'} = -\mu_T \, \partial_z \overline{u}$$

and

$$\tau_0 = (\mu + \mu_T) \, \partial_z \overline{u}. \tag{14.30}$$

Then the viscous sublayer may be defined as the region, where it is possible to assume $\mu_T = 0$. Accordingly, the logarithmic boundary layer is the area where $\mu \ll \mu_T$. Comparing (14.30) and (14.28) in this case we obtain the expression for the turbulent viscosity, namely:

$$\partial_z \overline{u} = \frac{\tau_0}{\mu_T} = \frac{v_*^2 \rho}{\mu_T} \sim \frac{v_*}{z}.$$

Thus

$$\mu_T = \kappa \rho v_* z,$$

and the greater is the distance from the wall (z) and the shear stress (v_*), the more intense is mixing.

14.3.4. Influence of Roughness. The Roughness Parameter

Consider now a rough wall in the case when the roughness height h is comparable with the value of the scale $z_* = \dfrac{\nu}{v_*}$. On the one hand, such protrusions, their height, shape and location should influence the flow. However, on the other hand, it seems obvious that far from the wall the impact of protrusions is averaged, integral. Therefore, the type of dependence (14.29), is likely to remain the same, *i.e.*

$$\frac{\overline{u}}{U} = \ln\frac{z}{z_0},$$

but the values of parameters U and z_0 will be determined by the nature of the surface roughness. Because the velocity scale U is determined by component of the momentum flux density τ_0, it may be assumed that

$$U = v_* = \sqrt{\frac{\tau_0}{\rho}}.$$

Thus, for the mean velocity profile near the rough wall we have

$$\overline{u} = v_*\ln\frac{z}{z_0}. \tag{14.31}$$

The unknown quantity z_0 substantially depends on the wall roughness and is the parameter which we use to describe the averaged interaction of the flow with roughness elements. It is called the *roughness parameter*. In accordance with (14.31) it turns out that the distance $z = z_0$ is the height where the mean velocity $\overline{u}(z_0)$ vanish if the logarithmic dependence would work at arbitrary distance from the wall.

NOTES

[1]Prandtl Ludwig (1875–1953), a German hydromechanics.

[2]Kármán, Theodor von (1881–1963), a German scientist in the field of mechanics.

Part III
Supplement

CHAPTER 15

Fluid Mechanics from an Observer's Viewpoint

Abstract: In this chapter[1] we focus our attention on one of the possible improvements of the classical model of the viscous fluid. The goal is to overcome the basic shortcoming of the classical model: the contradiction with the causality principle. To solve the problem we take into account an observer, consider him as an element of the model and build the model of the fluid from his unique point of view. This approach appears to be rather general and allows application in related areas such as heat and mass transfer, signal propagation in a moving continuum and others.

Keywords: Body configuration, Body force, Causality principle, Conservation law, Contact force, Metric tensor, Observer, Signal speed, Space-time continuum, Thermodynamics, Total energy, Viscous fluid model, World-line.

15.1. INTRODUCTION

The goal of developing of any physical theory and corresponding model of a physical phenomenon, is twofold. On the one hand, the model is constructed to predict possible evolution of characteristics of the phenomenon. On the other hand, this model should explain the phenomenon, *i.e.*, it should describe the phenomenon in accordance with the contemporary scientific notions.

The quality of the model can also be understood in two ways. In the first case, estimation of the quality of the model mainly consists in comparison of results of the modeling with observations. Which method is used to obtain these results is of minor importance, provided that it guarantee the efficiency of simulation (the accuracy and speed of receipt of the forecast). In the second case, on the contrary, efficiency recedes into the background and at the forefront in the estimation of the quality of the model is its adequacy, that is, the self-consistency of description and its compliance with most fundamental principles of our perception of the world. And first of all the laws of logic and causality.

Both of the above-mentioned viewpoints show that there are cases when the classical fluid model needs improvements. Firstly, the classical model of viscous, heat-conducting fluid cannot be considered as an adequate description, since it allows infinite velocities of propagation of disturbances and, thus, does not satisfy the causality principle. Secondly, the same reason prevents the use of this model in the relativistic context (*e.g.*, in astrophysics, cosmology, *etc.*), where the continuity hypothesis is applicable, but infinite velocities are unacceptable by virtue of the underlying postulates. Besides, there are certain situations, when solutions of the corresponding parabolic equations of the model fail to describe the experimental data. This mainly concerns the experimental studies of finite-speed propagation of the thermal signals at low temperatures as well as laser and microwave heating with extremely short duration or high frequencies.

Beginning from the forties of the last century numerous papers appear devoted to development of the causal fluid models. Nowadays there is a significant literature associated with this field of research.

As an example of possible approach to exclusion of the above-mentioned contradiction, here we briefly consider a suggested recently variant of the causal model of the viscous fluid (see [11-13] for details and references therein). It is based on an obvious fact that any model of a physical phenomenon needs information about the phenomenon which reaches the observer with some finite velocity. Hence, an adequate model should incorporate an observer and velocity of an information carrier, *i.e.*, it should describe an observation process. It may be shown that such a model prevents onset of velocities exceeding the signal speed and thus, hinders the arising of infinite velocities.

15.2. CAUSAL FLUID MODEL

15.2.1. General Considerations

Observations are an integral part of the process of knowledge. A theory of a phenomenon as a rule is based on information that reaches an observer using some kind of a carrier. If such a carrier is a radiation, we often call it signal.

If we intend to introduce the observation process into the theory, we are to consider the space of events (the four-dimensional space-time continuum) and associate each point of the body in study with its world line, *i.e.*, with the sequence of events connected with this point. It is easy to show that a vector $\vec{u}$ tangent to a world line is a four-dimensional velocity vector. Its 0-th component[2] may be interpreted as the speed of the signal used for observations while the rest three components are the components of the usual three-dimensional velocity vector of the point. Following this interpretation, the quantity $\frac{1}{2}|\vec{u}|^2$ is called the four-dimensional specific kinetic energy density.

To calculate this quantity it is necessary to choose a metric tensor g, since $|\vec{u}|^2 = g(\vec{u}, \vec{u})$. The components of the metric tensor are scalar products of basis vectors $g_{\alpha\beta} = g(\vec{e}_\alpha, \vec{e}_\beta)$. Hereafter, the orthogonal coordinate basis is used, when $g_{\alpha\beta} \propto \delta_{\alpha\beta}$. The coefficients of proportionality are scale factors which may be chosen to be equal to each other. In this case $g = g_0 I$, where I is the unit tensor, and the choice of the metric coefficient g_0 defines the metric tensor. To narrow the choice an additional requirement may be used. Here, we assume that $|\vec{u}| = 1$. This is possible since the signal speed (the 0-th component of the velocity vector) is always non-zero. Surely, the metric tensor may be chosen differently. However, the connection between the mass and energy of the body (which will be considered further) most probably would be hidden.

15.2.2. Basic Notions

We suppose that the continuity hypothesis is accepted and any object B is regarded as an infinitely divisible set of points every two of which have non-overlapping vicinities. Each point of the body B is associated with a set of events in the four-dimensional space-time continuum W or the space of events. These events form the world-line λ of this point of the body. The world-lines of all points of the body in total form the world-tube of the body B which is regarded to be a four-dimensional manifold B^4 in the space W.

World lines may be smoothly parameterized with a real parameter. Parameterization is arbitrary, and to narrow the choice additional restrictions may be imposed. The purpose is to leave only one parameter called time t, instead of infinite number of independent parameters of different world lines. This is achieved by allocating one world line called the world line of an observer λ_0, which parameterization is left arbitrary. The remaining parameters are synchronized with the time of the observer. The synchronization method is usually chosen such that it may be interpreted in terms of the signal speed. The phase speed of the signal hereafter is denoted by **s**. Therefore, the observer is the totality ($\lambda_0, t,$ **s**).

We emphasize that the letter s does not necessarily designates the speed of light. This is the speed of a signal used for observations, and this signal depends on a particular observation or experiment. Usually this signal is light, although sometimes it may be sound or something else.

Presence of an observer allows definition of spaces of simultaneous events W_t. A simultaneous cross-section B_t of the world-tube B^4 is called the configuration of the body at time t. A totality of configurations B_t of the body B is studied.

15.2.3. Conservation Laws

INTEGRAL CONSERVATION LAWS

Let us define measures $\mathcal{M}(\mathcal{B})$ and $\mathcal{V}(\mathcal{W})$ on the body $\mathcal{B}$ and the space of events $\mathcal{W}$, respectively. Both of them induce measures *Mass* $(\mathcal{B}_t)$ and $\mathcal{V}(\mathcal{B}_t)$ on the world-tube cross-section which are usually interpreted as the mass and volume of the configuration $\mathcal{B}_t$. The conservation law

$$d_t Mass(\mathcal{B}_t) = 0 \qquad\qquad (15.1)$$

is postulated.

It is possible to define another measure *Energy* on the same configuration $\mathcal{B}_t$ such that

$$Energy(\mathcal{B}_t) = Const \cdot Mass(\mathcal{B}_t), \quad Const = const.$$

This leads to another conservation law of *Energy* of the cross-section $\mathcal{B}_t$

$$d_t Energy(\mathcal{B}_t) = 0, \qquad\qquad (15.2)$$

which follows from of the conservation law (15.1). The interpretation of *Energy* is considered below.

DIFFERENTIAL CONSERVATION LAWS

From quite general assumptions it follows the existence of a unique function ρ, which binds the measure *Mass* on $\mathcal{B}_t$ with the volume $V(\mathcal{B}_t)$

$$Mass(\mathcal{B}_t) = \int_{V(\mathcal{B}_t)} \rho dV.$$

The measure *Energy* and the volume $V(\mathcal{B}_t)$, in turn, satisfy a similar relation

$$Energy(\mathcal{B}_t) = \int_{V(\mathcal{B}_t)} \kappa dV.$$

Both quantities ρ and κ are called densities of the mass and energy, respectively, and $\kappa = Const\ \rho$.

The following differential conservation laws

$$div(\rho\vec{u}) = 0 \qquad\qquad (15.3)$$

and

$$div(\kappa\vec{u}) = 0, \qquad\qquad (15.4)$$

correspond to equations (15.1) and (15.2). Here $\vec{u} = d_t$ is a velocity vector tangent to a world-line. Using (15.3), the latter equation may be written as follows

$$d_t \frac{\kappa}{\rho} = 0. \qquad\qquad (15.5)$$

Since κ is equal to $\frac{1}{2}|\vec{u}|^2\rho$ by definition and $\kappa = Const\ \rho$, one has $Const = \frac{1}{2}|\vec{u}|^2$. Choosing the metric tensor $g = g_0 I$ and the length of the velocity vector $|\vec{u}| = 1$ one finds $Const = \frac{1}{2}$ and $|\vec{u}|^2 = g(\vec{u}, \vec{u}) = g_0 I(\vec{u}, \vec{u}) = 1$ Thus,

$$g_0 = \frac{1}{I(\vec{u},\vec{u})} = \frac{1}{\sum_\alpha (u^\alpha)^2} \qquad\qquad (15.6)$$

Here u^α are components of the vector $\vec{u} = u^\alpha \vec{e}_\alpha$ with respect to the basis[3] $\{\vec{e}_\alpha\}_{\alpha=0}^3$. The index summation convention is used. From now on the Latin indices take values from 1 to 3 and the Greek indices take values from 0 to 3.

15.2.4. Thermodynamics and Total Energy Conservation Law

Since the continuity hypothesis is accepted, we are faced with the necessity of the scale separation of motion. This is done using the averaging and consideration of the so-called thermodynamic case. The averaging procedure divides each world-line into two components: the first one is smooth, while another is fluctuating. The corresponding tangent vector field $\vec{u}$ also divides into the averaged vector field $\overline{\vec{u}}$ which vectors are tangent to the smoothed world-lines, and fluctuating vector field $\vec{u}'$

$$\vec{u} = \overline{\vec{u}} + \vec{u}'.$$

Consider the length of velocity vector $1 = |\vec{u}|^2 = |\overline{\vec{u}} + \vec{u}'|^2$ Denoting $\overline{k} = \frac{1}{2}\rho|\overline{\vec{u}}|^2$ and $e = \frac{1}{2}\rho\left(|\vec{u}|^2 - |\overline{\vec{u}}|^2\right)$ we define thereby densities of the kinetic energy of mean motion and the internal energy, respectively. Such division of energy leads to redefinition of the metric coefficient g_0, which now is written as follows (cf. formula (15.6))

$$g_0 = \frac{1-2e/\rho}{\Sigma_\alpha (u^\alpha)^2}. \tag{15.7}$$

Henceforth, we drop the overbar notation and will think of k and $\vec{u}$ as already averaged quantities.

As has been mentioned above, the quantity *Energy* $(\mathcal{B}_t)$ may be regarded as the kinetic energy of a cross-section of the world-tube. Averaging divides this quantity into two parts: $K(\mathcal{B}_t)$, the kinetic energy of mean motion, and $E(\mathcal{B}_t)$, the internal energy of a cross-section of the world-tube

$$Energy(\mathcal{B}_t) = K(\mathcal{B}_t) + E(\mathcal{B}_t), \quad K, E > 0.$$

The left side is then called the *total energy* of the configuration $\mathcal{B}_t$. The derivatives of K and E with respect to volume $V(\mathcal{B}_t)$ are the above-introduced densities k and e, respectively. In the thermodynamic case, equation (15.3) does not change and the equation (15.4) turns into the total energy $(k+e)$ conservation law

$$div(k + e)\vec{v} = \rho d_t \frac{k+e}{\rho} = 0. \tag{15.8}$$

15.2.5. The Equations of Momentum and Internal Energy Balance

To obtain the equation of momentum balance we write equation (15.4) in terms of the momentum flux density. The tensor $M = \rho \vec{u} \otimes \vec{u}$ is called the density of the 4-*momentum flux*. The continuity equation (15.3) may be used to show that (15.8) is equivalent to the equation

$$g(\vec{u}, div M) + \rho d_t \frac{e}{\rho} = 0. \tag{15.9}$$

The choice of expressions for components of div M defines the model of continuum. For this purpose the *stress tensor* T is introduced and the so-called *equation of motion* or the momentum balance equation

$$div M = div T \tag{15.10}$$

is postulated. Substituting (15.10) in (15.9) one obtains the internal energy balance equation

$$\rho d_t \frac{e}{\rho} = -g(\vec{u}, div T). \tag{15.11}$$

The system of equations (15.3, 15.10 and 15.11) together with the equation of state and definition of the tensor T is the causal model of a fluid.

15.2.6. Definition of the Stress Tensor

We construct the stress tensor with the help of the velocity gradient and the metric tensor.

If the stress tensor T is set proportional to the metric tensor

$$T = \pi g^{-1}, \tag{15.12}$$

then the fluid is called *Perfect* or *Ideal*. For $g = g_0^{-1} I$ the stress tensor T is diagonal. The quantity π is a function of coordinates. Its geometrical interpretation will be discussed in the Section 15.4.3.

When the stress tensor contains a correction term which is proportional to the symmetric part of the velocity gradient $\nabla \vec{u}$, known as *the deformation rate tensor D*

$$T = \pi g^{-1} + 2\eta D, \quad 2D^{\alpha\beta} = g^{\alpha\gamma} \nabla_\gamma u^\beta + g^{\beta\gamma} \nabla_\gamma u^\alpha,$$

the fluid model is called *Viscous*. Here ∇_γ is the covariant derivative in the direction of $\vec{e}_\gamma$. This additional term is called the *viscous stress tensor*. When $g = g_0 I$, the stress tensor simplifies to the form

$$T^{\alpha\beta} = \pi g_0^{-1} \delta^{\alpha\beta} + \eta g_0^{-1} \left(\delta^{\alpha\gamma} \partial_\gamma u^\beta + \delta^{\beta\gamma} \partial_\gamma u^\alpha \right). \tag{15.13}$$

Here, $\delta^{\alpha\beta}$ denotes a component of the identity tensor I and $\partial_\gamma = \partial_{x\gamma}$.

15.3. STANDARD FLUID MODEL

For the purpose of comparing the causal model with the standard one here we briefly repeat derivation of equations of the standard model[4], using causal formalism. In the standard case the body is just the same object as in the previous consideration. It is associated with its world-tube, the four-dimensional manifold $\mathcal{B}^4$, in the space of events $\mathcal{W}$.

Space-time continuum $\mathcal{W}$ is regarded to be $\mathcal{W} = \mathcal{T} \times \mathcal{L}$ where $\mathcal{T}$ is a 1D space of time instants and $\mathcal{L}$ is a 3D space of places. The mapping $\phi : \mathcal{W} \to \mathbb{R}^1 \times \mathbb{R}^3$ is called the frame of reference. Although one may associate an observer with the origin of the frame of reference, the standard fluid mechanics does not include it in the theory. All possible observers are equal to each other and their spaces of simultaneous events are the same. In the standard case time and space have the absolute meaning and this may be interpreted as infinity of the signal velocity. In fact, the absolute time plays the role of a parameter of the world-lines.

The frame of reference ϕ provides each event with four numbers (t, x^1, x^2, x^3), where the first one is the time of the event, and the rest three are the coordinates of a place. If the numbers (x^1, x^2, x^3) are associated with a point of the fluid body then they are called the Euler coordinates of this point at time t. The functions $(t, x^1(t), x^2(t), x^3(t))$ give a parametric description of the world-line of the point of the body.

Parameterization of the world-lines generates a tangent vector field on the manifold $\mathcal{B}^4$. If $\{\vec{e}_\alpha\}_{\alpha=0}^3$, $\vec{e}_0 = \partial_t$, $\vec{e}_k = \partial_{x^k}$, $k = 1,2,3$ is a coordinate basis defined at each point of the world-tube $\mathcal{B}^4$, then $v = v^\alpha e_\alpha$ and $v^0 = 1$.

15.3.1. Measures and Mass Conservation Law

All considerations concerning measures and conservation of mass remain unchanged in the standard theory. Meanwhile the resulting equations are different, since the tangent velocity vector within the framework of the standard fluid mechanics has three non-constant components only $\vec{u} = (1,u^1,u^2,u^3) = (1,\vec{v})$, where $\vec{v} = (v^1,v^2,v^3)$ is the three-dimensional vector and $u^i = v^i$. Thus $div\ \vec{v} = div\ \vec{u}$ and the continuity equation is as follows

$$\partial_t \rho + div(\rho \vec{v}) = 0. \tag{15.14}$$

15.3.2. Metric Tensor and the Kinetic Energy Conservation Law

At first sight four-dimensional reasoning is also possible here. For instance, if we define the metric tensor such that the velocity vector length is constant, the conservation of energy would be then a necessary consequence of the conservation of mass. However, such an approach meets difficulties since the dimensional homogeneity of the space of events in the standard case is absent. In order to construct a dimensionally homogeneous space of events it would be necessary to redefine time coordinate using an arbitrary constant with physical dimension of velocity. Since the time coordinate is imaginary (as in the previous consideration) or the metric tensor of the space of events is indefinite) this velocity actually is the upper limit for observed three-dimensional velocities (see [11] for discussion). And such limit is arbitrary.

Thus, we need to choose this constant greater than any possible value of velocity. Inasmuch as the standard fluid mechanics has no natural upper velocity limit, it must be chosen infinite. The metric coefficient (15.6) becomes zero and another metrics should be used. In case of Cartesian coordinates the metric tensor g_3 is chosen to be the unit tensor I. Further for simplicity sake we shall discuss this case only.

Due to new definition of the metric tensor the length of the velocity vector no longer remains constant. The kinetic energy conservation does not follow from the mass conservation law and should be postulated instead.

15.3.3. Thermodynamics and the Total Energy Conservation Law

After the continuity hypothesis is accepted, the explicit description is applicable only to large-scale motions. In this case we are forced to introduce the so-called three dimensional internal energy $E_3(\mathcal{B}_t)$ of the body and substitute the postulated energy conservation law for the total energy conservation law $d_t(K_3 + E_3) = 0$, where K_3 is the three dimensional kinetic energy of the body. Corresponding differential equation is as follows

$$d_t \frac{k_3 + e_3}{\rho} = 0. \tag{15.15}$$

Here $k_3 = \frac{1}{2}\rho g_3(\vec{u}, \vec{u}) = \frac{1}{2}\rho \delta_{jk} v^j v^k$, g_3 is the metric tensor and

$$K_3(\mathcal{B}_t) = \int_{V(\mathcal{B}_t)} k_3 dV, \qquad E_3(\mathcal{B}_t) = \int_{V(\mathcal{B}_t)} e_3 dV$$

When the force power W and the external heating Q are known, the total energy balance takes the form $d_t(K_3 + E_3) = W + Q$. This equation is called the first law of thermodynamics.

15.3.4. The Momentum and Internal Energy Balance Equations

Let $\mathbf{m}(\mathcal{B}_t) = \int_{V(\mathcal{B}_t)} \rho \vec{v} dV$ be a momentum of the body at time t and f be a force which acts on the body. Then the basic principle of the fluid mechanics reads

$$d_t \mathbf{m}(\mathcal{B}_t) = f. \tag{15.16}$$

Derivation of the differential equation of the momentum balance requires expressing the force f in terms of the stress tensor $f = \int_{V(\mathcal{B}_t)} div T_3 dV$ (for simplicity we do not consider the body forces here). This gives

$$\rho d_t \vec{v} = \partial_t \rho \vec{v} + div M_3 = div T_3. \tag{15.17}$$

Here M_3 is the 3D momentum flux density tensor using the equation (15.15), relation $d_t \frac{k_3}{\rho} = g_3(\vec{v}, d_t \vec{v}) = \delta_{jk} v^j d_t v^k$ and the local thermodynamic equilibrium hypothesis we obtain the following balance equations

$$\rho d_t \frac{k_3}{\rho} = g_3(\vec{v}, div M_3) = g_3(\vec{v}, div T_3),$$

(15.18)

$$\rho d_t \frac{e_3}{\rho} = T_3 : \nabla \vec{v}.$$

Here the inner product of two tensors gives

$$T_3 : \nabla \vec{v} = tr(T_3(\nabla \vec{v})) = (T_3)^j_k (\nabla \vec{v})^k_j.$$

In the standard case the system of equations of fluid mechanics consists of equations (15.14), (15.17) and (15.18) together with the equation of state and corresponding choice of the stress tensor.

15.3.5. Definition of the Stress Tensor

The stress tensor is constructed using the metric tensor g_3 and the velocity gradient $\nabla \vec{v}$. Note, that here we assume that $g_3 = I$.

Perfect fluid model implies that the stress tensor T_3 is proportional to the metric tensor

$$T_3 = -p_3 g_3^{-1} = -p_3 I.$$

(15.19)

The quantity p_3 is a function of coordinates and is called *pressure*. The equation of motion for the perfect fluid is written as

$$\rho d_t \vec{u} = -\nabla p_3.$$

(15.20)

Viscous fluid model implies that the stress tensor contains a correction term dependent on *the deformation rate tensor* D_3 equal to the symmetric part of the velocity gradient $\nabla \vec{u}$

$$T_3 = -p_3 g_3^{-1} + 2\mu D_3, \qquad 2(D_3)^{jk} = (g_3)^{jn} \nabla_n v^k + (g_3)^{kn} \nabla_n v^j.$$

The correction term is called the *viscous stress tensor* and the factor μ is known as the *dynamic viscosity*. Due to the choice of the metric tensor the stress tensor for the viscous fluid takes the form

$$(T_3)^{jk} = -p_3 \delta^{jk} + \mu \left(\delta^{jl} \partial_l v^k + \delta^{kl} \partial_l v^j \right). \tag{15.21}$$

15.4. COMPARISON OF THE STANDARD AND CAUSAL FLUID MODELS

15.4.1. Standard Model as a Limit Case for the Causal Model

Two main assumptions underlie the standard fluid model: the absoluteness of time and the local thermodynamic equilibrium. The causal theory, on the contrary, does not need none of them. Inasmuch as the first assumption is interpreted as infinity of the signal velocity, we should be able to derive standard equations *via* passage to limit of the causal model. This possibility will be illustrated for the case of the perfect fluid (although the general situation may also be considered).

Please note, that physical dimensions of some quantities used in both theories (causal and standard) are different (see Table **15.1**). Let the units of mass, length and time are denoted by M, L, T and $[a]$ is a dimension of a quantity a. Due to definition of the metrics in the causal theory the length of the velocity vector is non-dimensional and dimension of components of the metric tensor is as follows $[g] = T^2 L^{-2}$. The dimension of densities of kinetic and internal energies is equal to dimension of the mass density: $[\rho] = [k] = [e] = [\pi] = ML^{-3}$. The dimension of components of the momentum flux density tensor, stress tensor and the pressure is usual: $[M] = [T] - [p] = ML^{-1}T^{-2}$.

Table 15.1. Causal theory *vs.* standard: physical dimensionsamerican.

Causal Theory	Std Theory	Causal Theory	Standard Theory
$[g] = T^2 L^{-2}$	$[g_3] = 1$	$[M] = [T] = ML^{-1}T^{-2}$	$[M_3] = [T_3] = ML^{-1}T^{-2}$
$[\rho] = ML^{-3}$	$[\rho] = ML^{-3}$	$[k] = [e] = [\pi] = ML^{-3}$	$[k_3] = [e_3] = [p_3] = ML^{-1}T^{-2}$

The initial four-dimensional system consists of equations (15.3), (15.10) and (15.11) and the resulting system includes equations (15.14), (15.17) and (15.18). All considerations are carried out using the so-called normal coordinates (see, *e.g.*, [5]), which permit local substitution of covariant derivatives for partial ones.

Pressure. The four-dimensional stress tensor $T = \pi g^{-1}$ with respect to the orthogonal basis is proportional to the unit tensor $T = \pi g_0^{-1} I$. The pressure in the standard fluid mechanics and the scale factor πg_0^{-1} in the causal theory play similar roles and therefore may be associated with each other. We shall write $\pi g_0^{-1} = -p$, but it should be kept in mind that the standard pressure is equal to

$$p_3 = lim_{s \to \infty} p.$$

The momentum balance equation. Consider four-dimensional momentum balance equation (15.10). Using the continuity equation (15.3), this may be written as follows

$$\rho d_t \vec{u} = divT = -div(pI),$$

or in the coordinate form

$$\rho d_t u^\beta = \rho\big(\partial_t u^\beta + u^k \partial_k u^\beta\big) = -\partial_\alpha p \delta^{\alpha\beta}. \tag{15.22}$$

For $\beta = k$ equation (15.22) gives the known three-dimensional Euler equations with respect to the three-dimensional velocity $\vec{v} = (v^1, v^2, v^3)$. For $\beta = 0$ and a limited rate of the pressure change we find

$$\rho d_t u^0 = -\partial_\alpha p \delta^{\alpha 0} \quad \Rightarrow \quad d_t \mathbf{s} = \frac{1}{\rho \mathbf{s}} \partial_t p \quad \Rightarrow \quad \lim_{s \to \infty} d_t \mathbf{s} = 0.$$

The continuity equation. Now consider the four-dimensional continuity equation (15.3)

$$0 = \partial_\alpha(\rho u^\alpha) = \rho(\partial_0 u^0 + \partial_k u^k) + u^0 \partial_0 \rho + u^k \partial_k \rho \tag{15.23}$$

$$= \rho \left(\frac{1}{\mathbf{s}} \partial_t \mathbf{s} + \partial_k v^k\right) + \partial_t \rho + u^k \partial_k \rho \tag{15.24}$$
$$= \frac{\rho}{\mathbf{s}} \partial_t \mathbf{s} + d_t \rho + \rho \partial_k u^k.$$

When $\mathbf{s} \to \infty$ the first term in (15.23) tends to zero (since $d_t \mathbf{s} \to 0$; see the previous item) and the rest part is just the standard continuity equation.

The internal energy balance equation. At last, let us consider the four-dimensional internal energy balance equation (15.11). In case of the perfect fluid we get

$$\rho d_t \frac{e}{\rho g_0} = -\delta_{\alpha\beta} u^\beta \nabla_\gamma \frac{\pi}{g_0} \delta^{\alpha\gamma} = d_t p.$$

Denoting $e_3 \equiv \lim_{s\to\infty} e g_0^{-1}$ the standard internal energy and using the hypothesis of local thermodynamic equilibrium which means that $d_t \frac{p}{\rho} = 0$, we obtain the standard internal energy balance equation

$$d_t \frac{e_3}{\rho} = \frac{1}{\rho} d_t p_3 = d_t \frac{p_3}{\rho} - p_3 d_t \frac{1}{\rho} = -p_3 d_t \frac{1}{\rho}.$$

15.4.2. Basic Definitions, Axioms and Theorems

Now we may briefly compare the causal fluid model with the standard one. Both theories differ from each other in resulting equations, in a number of basic definitions, axioms and theorems. In what follows the letters D, A and T denote Definition, Axiom and Theorem, respectively.

Table 15.2. Causal theory *vs.* standard: general definitions .

Causal Theory	Std Theory	Causal Theory	Std Theory
D1. time t		D5. metric tensor fields in $\mathcal{W}$ and $\mathcal{L}$	
relative	absolute	g	g_3
D2. velocity vector		D6. kinetic K and internal E energy densities	
$\vec{u} = (u^0, u^1, u^3, u^4)$	$\vec{v} = (v^1, v^3, v^4)$, $v^i = u^i$	$k = \frac{1}{2} \rho g(\vec{u}, \vec{u})$	$k_3 = \frac{1}{2} \rho g_3(\vec{v}, \vec{v})$
D3. configuration of body $\mathcal{B}_t$		D7. momentum flux density	
relative	absolute	$M = \rho \vec{u} \otimes \vec{u}$	$M_3 = \rho \vec{v} \otimes \vec{v}$
D4. measures *Mass* and V on $\mathcal{B}_t$			

Note that other statements may be chosen for axioms. For instance, the axiom A2 in the standard theory is often regarded as a theorem, and the theorem T3 is instead treated as an axiom, which is known (together with the local thermodynamic equilibrium hypothesis) as the first law of thermodynamics.

These tables show that in comparison with one axiom (A1) in the causal model (conservation of measure), there are two axioms (A1 and A2) in the standard model. This is due to the fact that in the causal model the mass and total energy are two interpretations of measure defined on $\mathcal{B}_t$, while in the standard theory these quantities are independent measures. However, we need the same number of axioms as in the standard model[5], to make up the closed system of causal differential equations. Whereas in the standard case the definition of the stress tensor is regarded as some additional assumption, in the causal case the specification of T should be considered as the lacking axiom which in the end determines what portion of the total energy is treated as the internal energy.

Table 15.3. Causal theory *vs.* standard: main axioms and theorems.

Causal Theory	Std Theory	Causal Theory	Std Theory
A1. mass conservation: $d_t Mass = 0$		T2. kinetic energy balance:	
$(\nabla, \rho\vec{u}) = 0$	$\partial_t \rho + (\nabla, \rho\vec{v}) = 0$	$\rho d_t \dfrac{k}{\rho} = g(\vec{u}, div M)$	$\rho d_t \dfrac{k_3}{\rho} = g_3(\vec{v}, div M_3)$
total energy conservation:		T3. internal energy balance:	
T1. $d_t \dfrac{k+e}{\rho} = 0$	A2. $d_t \dfrac{k_3+e_3}{\rho} = 0$	$\rho d_t \dfrac{e}{\rho} = -g(\vec{u}, div T)$	$\rho d_t \dfrac{e_3}{\rho} = -g_3(\vec{v}, div T_3)$
D8. momentum balance (tensor T definition):			
$div M = div T$	$\partial_t \rho\vec{v} + div M_3 = div T_3$		

15.4.3. Some Parallels Between Basic Notions of the Fluid Models

Classical notion of time and the world-line parameter. Unlike the standard fluid model in the causal fluid mechanics time is defined only by an observer. Different observers introduce their own times all of which may be used for constructing the theory. Such understanding of time being much more physically justified, deeply unites the whole theory, all parts of which become closely connected with each other.

Fluid body and its configuration. New understanding of time leads to redefinition of notion of configuration of a body. The standard theory regards configuration of a body as a function of time only. In the causal theory each configuration depends also on the observer, its position and signal speed. Indeed, different observers have different spaces of simultaneous events. Cross-sections of the world-tube of the body, in their turn, depend on location of the body as well as its displacement relative to the observer. Besides, these cross-sections together with the condition which makes the body observable, are defined, to a certain extent, by the speed of the signal chosen by the observer.

Forces and curvature of the world-lines. When a world-line of a point of a body is non-parallel to the world-line of the observer this means that this point moves with respect to the observer. If this world-line differs from the straight line then the motion is non-uniform. Consider factors which are responsible for such non-uniform motion. We call the quantity x_*

$$x_* = \frac{\int_{V(\mathcal{B}_t)} \rho x dV}{\int_{V(\mathcal{B}_t)} \rho dV} = \frac{1}{Mass(\mathcal{B}_t)} \int_{V(\mathcal{B}_t)} \rho x dV.$$

the centre of mass of the configuration $\mathcal{B}_t$. The curve $x_*(t)$ is the world-line of the centre of mass. Its curvature characterizes the mean curvature of the world-tube of the body and to some extent the curvatures of the world-lines.

The factors which cause incurvation of the world-line of the centre of mass will be called the *external* factors. The change of the frame of reference which straightens the world-line of the centre of mass annuls the influence of external factors on all of the world lines of the points of the body. Other factors which bend the world-lines of the points of the body leaving the world-line of the centre of mass unaffected we shall call *internal*.

The curvature of the world-lines which is connected with acceleration of a moving point, is traditionally explained by the body–environment interaction and is described using the notion of the body forces f_B and surface forces f_S. The main principle of dynamics in terms of forces reads

$$d_t(Mass \, d_t x_*) = f_B + f_S \tag{15.25}$$

or

$$\int_{V(\mathcal{B}_t)} \rho d_t \vec{u} dV = f_B + f_S$$

(cf. this expression with equation 15.16). The quantity in parentheses in the left-hand side of (15.25) is called the momentum of the body $\mathcal{B}$ and may be regarded as a vector tangent to the curve $x_*(t)$. The surface and body forces are associated with the internal and external factors, respectively.

The internal factors (surface forces) are taken into account with the help of the so-called stress tensor T. Both theories use similar definition of this tensor with the only difference that a standard stress tensor is a 3×3 part of the causal one.

The model of the fluid is called perfect if the stress tensor is set to be proportional to the metric tensor

$$T = \pi g^{-1} = \pi g_0^{-1} I \equiv -pI. \tag{15.26}$$

In the standard theory the metric tensor g_3 is usually considered as the unit tensor. The function p, which is called pressure, is connected with the curvature of the space-time (see below).

Similar considerations may be applied in case of the viscous fluid when the stress tensor includes a correction term

$$T = \pi g^{-1} + 2\eta D = -pI + 2\mu g_0 D. \tag{15.27}$$

In both theories the tensor D, which is called the deformation rate tensor, is a symmetric part of the velocity gradient (four and three-dimensional, respectively) and $\mu \equiv g_0^{-1}\eta$ is known as the dynamic viscosity. Note that if g_0 is defined by (15.7) then $\eta < 0$, due to $g_0 < 0$ and $\mu > 0$.

The external factors (body forces) may be taken into account in three ways. One may specify an additional change in either

1. a kinetic energy: $k' = k + \rho\varphi(t)$, or

2. an acceleration value: $\rho d_t \vec{u} = div T + \rho \vec{b}$, or finally

3. a signal velocity: $(\mathbf{s}')^2 = \mathbf{s}^2 - \sigma^2$.

Pressure and the curvature of the space-time. To find out the geometrical meaning of the quantity π and associated with it pressure, consider a general case of the viscous fluid. Using definition (15.27) the equation (15.10) may be written as follows

$$\nabla_\alpha\left((M - 2\eta D)^{\alpha\beta} - g^{\alpha\beta}\pi\right) = 0. \tag{15.28}$$

If we interpret the symmetric tensor $\frac{1}{2}M - \eta D$ as the so-called Ricci tensor R, and the scalar π as the space-time scalar curvature R, then the previous equation becomes the well-known Bianchi identity for the Ricci tensor

$$\nabla_\alpha \left(R^{\alpha\beta} - \frac{1}{2} g^{\alpha\beta} R \right) = 0.$$

In the classical limit $\mathbf{s} \rightarrow \infty$ the diagonal components of the metric $g^{\alpha\alpha} = g_0^{-1} \rightarrow -\infty$ and since the pressure $p = -\pi g_0^{-1}$ is arbitrary, the scalar curvature $R \rightarrow 0$. This is the case of the standard fluid mechanics, the space-time of which is flat. On the contrary, the space-time of the causal fluid model is non-flat by definition and the pressure reflects, to some extent, the space-time incurvation.

15.5. CONCLUDING REMARKS

Problem and approach. The main goal of constructing causal theories is to satisfy the causality principle.

If we examine, for instance, derivation of the viscous fluid model, looking for a detail that was not taken into account and therefore caused contradiction with the causality principle, we find that such a detail is a process of observation, which is a prerequisite for a modelling, and which, however, is never incorporated into a fluid model. In the classical fluid mechanics it is always supposed that the observer immediately receives information about any phenomenon, although it is clear that information propagates with some finite speed (light, sound or whatever else). To make a model more realistic it is suggested to take observation process into account and to consider phenomena from the observer's viewpoint.

What are benefits of such an approach? Most important is that the signal speed appears to be the upper limit of resolved velocities in the fluid and thus it guarantees the causal stipulation of events. This limit, however, is not absolute, but a relative one, and makes sense within the current observation (or model). It does not forbid greater velocities, simply they will be wrongly interpreted by the observer. Any motion with greater velocity will be observed (if possible), as a motion with apparent velocity, which is smaller than the signal one (see [11]). Two models of an object, which differ in a signal used, must give different results, since real observations differ also.

What are possible signals and signal speeds? In most cases the light (or radio) signal is used. However, this is not the only possible signal. Many observations are carried out using sound signals (seismometry, echo-soundings, *etc*). Moreover, exceeding the bounds of contemporary physical experiment we discover a variety of possible information carriers and their velocities such as mail, eyewitness stories, verbal communication and the like. In due time, many useful environmental models have been constructed using information, which had been obtained this way. The lower the signal speed the lesser details are described by the model. A good example is the first model of the Gulf Stream, which was charted, using information gathered by sailors. Such model could predict rather generalized behavior of water masses particularly due to very low signal speed.

Besides, a variety of different signal speeds arise in the framework of numerical models of continuum.

Connection between mass and energy. It may be shown that the independence of mass and energy in the classical fluid mechanics is due to the choice of the metric tensor. In Cartesian case it is usually chosen to be equal to the unit tensor. By virtue of this choice the length of the velocity vector is non-constant and the kinetic energy conservation does not follow from the mass conservation law. It must be postulated instead, which is to be done.

Note that in both cases the energy is conserved due to either a postulate or a theorem.

The role of the observer. The main difference between standard and causal fluid models consists in the view on the role of the observer. While the standard fluid mechanics equalizes all observers, the causal approach, on the contrary, marks out a single observer and builds the fluid model from his unique viewpoint. All other potential observers are none the worse than the chosen one but they are different. It is impossible to mix them up since all of them have different pictures of the fluid motion. They have two distinguishing features: location and signal used for soundings of the continuum.

Two pictures of the fluid motion are different if they are built using signals with different speeds and/or different location of the observer. This is due to the fact that these features define the observer's frame of reference. Differences in these frames of reference give birth to different realizations of the spaces of simultaneous events and this leads to differences in simultaneous cross-sections of the world-tube of the body, *i.e,* its configurations. The faster the signal, the closer the configuration to the instantaneous one. In the standard case a configuration of the body is assumed to be instantaneous and pictures of the fluid flow obtained by various observers differ by translations only (leaving possible differences in scales and orientation of the spatial coordinate systems apart). The origin of this viewpoint is clear: the usual information carrier (light or radio waves) is an extremely fast signal.

Although the standard Navier-Stokes theory based on the idea of the infinite velocity of the signal propagation is suitable in many respects, some particular problems require the motion of a fluid be causally stipulated. Besides, if causality is assumed to be one of the most fundamental principles of the contemporary natural science the significance of causal theories becomes evident.

Thermodynamics. Causal theory does not utilize the local equilibrium hypothesis. For the sake of comparison with the standard fluid theories, such notions as temperature and entropy density have been redefined (see [12]). These new definitions have been constructed similar to standard definitions and such, that the basic idea remains unchanged: entropy of a perfect fluid is conserved in time. Thus, the difference in definitions is due to the difference in the form of the internal energy balance equations (Theorem 3 in the Table 2). While the standard equation is obtained using the local equilibrium hypothesis, the causal equation does not depend on it.

NOTES

[1]Parts of this chapter have been previously published in Acta Mechanica. December 2005, Volume 180, Issue 1, pp 83– 106. DOI: 10.1007/s00707-005-0262-y

[2]With respect to some basis, whose zeroth vector is co-directed with the time axis.

[3]Hereafter we use the coordinate bases. We define the so-called the observer's frame of reference, *i.e.* the map $\phi^t: \mathcal{W} \to i\mathbb{R}^1 \times \mathbb{R}^3$, where $i\mathbb{R}^1$ denotes the space of imaginary numbers. This frame of reference equips each point of $\mathcal{W}$ with four numbers $x = (x^0, x^1, x^2, x^3)$, the coordinates of the point. The imaginary coordinate x^0 is chosen such that $dx^0 = isdt$, where $i = \sqrt{-1}$ and s is the signal speed (see section 1.2.2). The components of vectors are further considered with respect to the coordinate basis $\{\vec{e}_\alpha\}_{\alpha=0}^3$, $\vec{e}_\alpha = \partial_{x^\alpha}$ chosen at each point of $u^\alpha = d_t x^\alpha$ and $u^0 = d_t x^0$.

[4]This model was discussed in the first part of the book.

[5]Note, that a model of a phenomenon is actually a totality of axioms. All other statements are theorems, *i.e.*, they are corollaries which follow from the indicated axioms.

CHAPTER 16

Memory — is in general a strange phenomenon. How hard is to memorize something, and how easy is to forget! Or that's what may be: you memorize one thing but recollect something quite different. Or: you memorize anything with difficulty, but very firmly, and later nothing can recollect. That also may happen. I would advise everyone to work a bit on their memory.

Daniil Harms
The Memoirs of a Wise Old Man. 1936-38.

Exercises

16.1. VECTOR FIELDS, TRAJECTORIES AND WORLD LINES

Solve the Cauchy problem

$$d_t \mathbf{x}(\mathbf{X}, t) = \vec{v}, \qquad \mathbf{x}|_{t=0} = \mathbf{X}$$

and find trajectories and world lines of the points of continuum.

1. One Dimensional Space of Places

Each point $\mathbf{x}$ of one-dimensional space of places is characterized by a single coordinate x and corresponding velocity vector $\vec{v}$ has only one component v. Any trajectory is described by a single equation $d_t x = v$ with the initial condition $x|_{t=0} = X$ where X is the Lagrangian coordinate of corresponding point of the body.

In all tasks you should do the following:

- find $x(X, t)$ together with $v(X, t)$ or $v(x, t)$
- calculate the Jacobian matrix and the Jacobian $J = |\partial_X x|$ of the coordinate transformation (of the Lagrangian coordinate $\mathbf{X}$ to the Euler coordinate $\mathbf{x}$)
- draw the velocity vector field tangent to trajectories of the points of the medium.
- draw the vector field of vectors tangent to the world lines and the world lines of the points of the medium. Finally draw trajectories which correspond to the world lines you have drawn.

Given :	*Answer* :		
a) $v(x, t) = t$,	$x = X + \frac{1}{2}t^2$,	$J = 1$,	
b) $v(x, t) = x$,	$x = Xe^t$,	$J = e^t$,	$v = Xe^t$,
c) $v(x, t) = xt$,	$x = Xe^{\frac{1}{2}t^2}$,	$J = e^{\frac{1}{2}t^2}$,	$v = Xte^{\frac{1}{2}t^2}$,
d) $v(X, t) = X + t$,	$X = \frac{x - \frac{1}{2}t^2}{t+1}$,	$J = t + 1$,	$v = \frac{x + \frac{1}{2}t^2 + t}{t+1}$.

Solutions:

Tasks a) – c) are carried out equally as follows (here, we consider the task a):

- Substituting the values of velocity in the right-hand side, we solve the Cauchy problem . (NB! the equality v=t means that velocity only numerically is equal to time).

$$d_t x = t, \quad x|_{t=0} = X.$$

Integrating, we obtain

$$\int_{x(0)}^{x(t)} dx = \int_0^t t\,dt \quad \Rightarrow \quad x|_X^x = \frac{1}{2}t^2|_0^t \quad \Rightarrow \quad x = X + \frac{1}{2}t^2.$$

- The Jacobi matrix consists here of one element $\partial_X x = 1$. Hence, $J = 1$.
- The steps of sketching of the velocity vector field, trajectories of the points of the medium, as well as the vector field tangent to the world lines and of the world lines of the points of the medium are shown in Fig. (**1**).

In the task g) the integration of the Cauchy problem gives

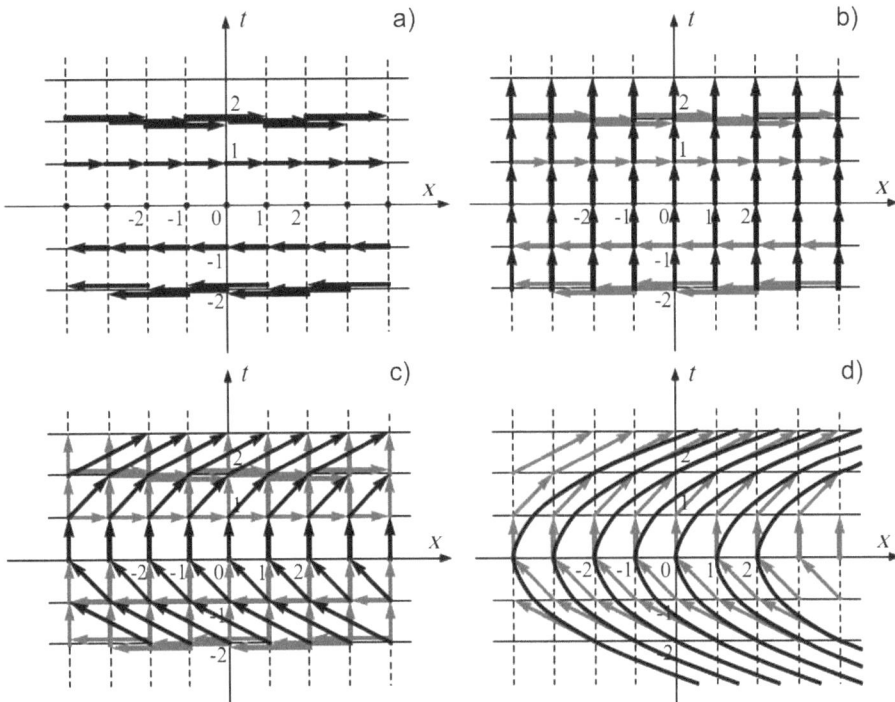

Fig. (1). Steps of sketching (coinciding vectors of the field v are slightly separated for clarity): a) the trajectories of the points of the medium (solid lines) and the velocity vector field, *i.e.*, vectors $v=d_t x$ tangent to the trajectories; b) the components of vectors $\vec{u}=(d_t t, d_t x)=(1,v)$ tangent to the world lines; c) vector field tangent to the world lines; d) the world lines of points of the medium.

$\int_{x(0)}^{x(t)} dx = \int_0^t (X + t)dt \quad \Rightarrow \quad x|_X^x = Xt + \frac{1}{2}t^2|_0^t \quad \Rightarrow \quad x = X(t + 1) + \frac{1}{2}t^2.$

Replacing the Lagrangian variable *via* its expression in terms of the Euler variables we find the velocity v as a function of time and place:

$$v = X + t = \frac{x - \frac{1}{2}t^2}{t + 1} + t = \frac{x + \frac{1}{2}t^2 + t}{t + 1}.$$

The rest is done as in tasks a) - c).

2. Two Dimensional Space of Places

In all tasks, do the following:

- find $\mathbf{x}(\mathbf{X}, t)$ and $\vec{v}\,(\mathbf{X}, t)$,
- calculate the Jacobian matrix and the Jacobian $J = |\partial_{X_j} x_k|$ of the coordinate transformation (of the Lagrangian coordinates $\mathbf{X}$ to the Euler coordinates $\mathbf{x}$)
- draw the velocity vector field and corresponding trajectories of the points of the medium.

Given :	*Answer* :	
a) $\vec{v}(\mathbf{x}, t) = x_1 \vec{e}_1,$	$\mathbf{x} = (X_1 e^t, X_2),$	$J = 1,$
b) $\vec{v}(\mathbf{x}, t) = \alpha x_1 \vec{e}_1 + \beta x_2 \vec{e}_2,$	$\mathbf{x} = (X_1 e^{\alpha t}, X_2 e^{\beta t}),$	$J = e^{(\alpha + \beta)t},$
c) $\vec{v}(\mathbf{x}, t) = -x_2 \vec{e}_1 + x_1 \vec{e}_2,$	$z = Z e^{it}, z \equiv (x_1 + i x_2),$	$J = 1.$

Solutions:

In the tasks a) and b) each equation (in vector notation) describing trajectories is equivalent to the systems of two independent equations, which are solved independently of one another, just as in the case of one-dimensional space of places. For example, in the task a) we have

$$\begin{cases} d_t x_1 &= x_1, \\ d_t x_2 &= 0. \end{cases}$$

The equation in the task c) is equivalent to a system of coupled equations

$$\begin{cases} d_t x_1 = -x_2, \\ d_t x_2 = x_1, \end{cases}$$

which may be easily solved with the transition to the complex plane. Firstly, we multiply, for example, the second equation by i and add it to the first equation. Next, we introduce the notation $z \equiv (x_1 + ix_2)$ and $Z \equiv (X_1 + iX_2)$ and obtain one equation with respect to the complex variable z: $dz = iz$ with the initial condition $z|_{t=0} = Z$. The solution of this Cauchy problem is $z = Ze^{it}$. Using the representation $e^{it} = \cos t + i\sin t$ we obtain

$$z = (X_1 + iX_2)(\cos t + i\sin t) = (X_1 \cos t - X_2 \sin t) + i(X_1 \sin t + X_2 \cos t).$$

Equating real and imaginary terms we have $x_1 = (X_1\cos t - X_2\sin t)$, $x_2 = (X_1\sin t + X_2\cos t)$. The Jacobi matrix and its determinant are as follows:

$$J = det \begin{pmatrix} \cos t & -\sin t \\ \sin t & \cos t \end{pmatrix} = 1.$$

Some representatives of the considered vector field and corresponding trajectories are shown in Fig. (*2).

16.2. PROPERTIES OF THE 2ND RANK TENSORS

1. Transposition

Compute the following tensors.

Note: In all cases the problem lies in the fact that such operations as transposition, summation, contraction, computation of inverse tensor (matrix), and so on, which were defined, say, for tensor A or B were not defined for tensors $(A + B)$, (A^T), *etc.* It is required to express these operations, using existing definitions. For example, the transposition of a tensor $(A + B)$ should be expressed in terms of the tensors A and/or B and operations which are already defined, like A^T, B^T, *etc.*

a. $(A + B)^T = ?$,
b. $(AB)^T = ?$,
c. $(A^T)^T = ?$,

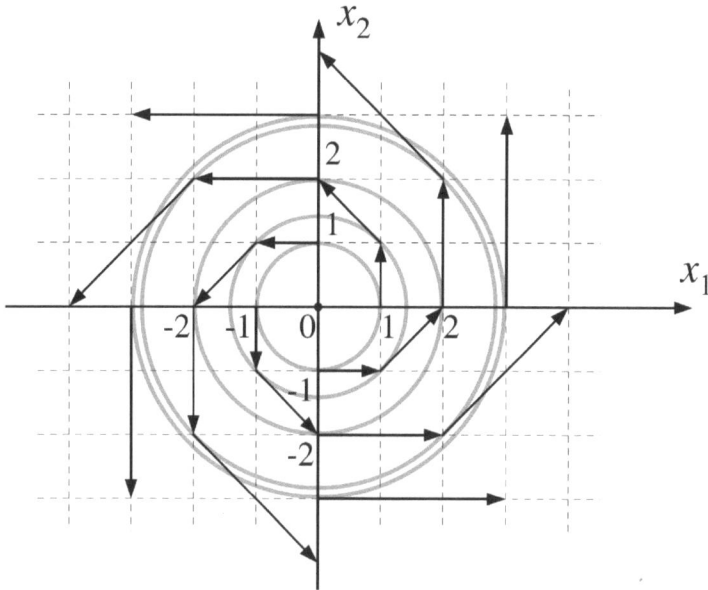

Fig. (*2). Tangent vector field and corresponding trajectories from the task 2c) on p.210.

Solutions. Here the general approach consists in considering a component of a tensor, (since, *e.g.*, transposition may be introduced or removed *via* changing the order of indices) and reducing the expression to already defined operations. Finally we try to restore the initial order of indices, so as to obtain the non-component form of the result.

 a. Firstly, we remove transposition (change the order of indices)

$$(A + B)^T_{jk} = (A + B)_{kj}.$$

Next, we use the definition of summation of tensors and write

$$(A + B)_{kj} = A_{kj} + B_{kj}.$$

And finally we restore the order of indices

$$A_{kj} + B_{kj} = A^T_{jk} + B^T_{jk}$$

and may write the answer in the indexless (tensor) form

$$(A + B)_{jk}^T = A_{jk}^T + B_{jk}^T \;\Rightarrow\; (A + B)^T = A^T + B^T.$$

b. Firstly, we remove transposition and compute the *kj*-component of the contraction

$$(AB)_{jk}^T = (AB)_{kj} = A_{kn}B_{nj}.$$

Next, we restore the original order of indices in accordance with the numbering convention for matrix elements and the rule of matrix multiplication.

$$A_{kn}B_{nj} = B_{nj}A_{kn} = B_{jn}^T A_{nk}^T = (B^T A^T)_{jk}.$$

Thus,

$$(AB)_{jk}^T = (B^T A^T)_{jk} \;\Rightarrow\; (AB)^T = B^T A^T.$$

c.
$$\left(A^T\right)_{jk}^T = A_{kj}^T = A_{jk} \;\Rightarrow\; \left(A^T\right)^T = A.$$

2. Inverse tensor

Compute the following tensors, assuming that A and B are reversible tensors.

a. $(A^{-1})^{-1} =?$
b. $(AB)^{-1} =?$
c. $(aA)^{-1} =?, a \neq 0.$
d. $(A^{-1})^T =?$

Solutions:

All reasonings are based on the definition of the inverse tensor $T^{-1} T = I$. Firstly, considering the given tensor as, say, T^{-1} we construct the latter equality. Then multiplying both sides by this or that tensor and using the same equality we try to leave on the left side the tensor T^{-1} only.

(a) $(A^{-1})^{-1}(A^{-1}) = I \;\Rightarrow\;$ multiply from the right by A
$$\Rightarrow\; (A^{-1})^{-1}A^{-1}A = A \;\Rightarrow\; (A^{-1})^{-1} = A.$$

(b) $(AB)^{-1}AB = I \;\Rightarrow\;$
$$\Rightarrow\; (AB)^{-1}A = B^{-1} \;\Rightarrow\; (AB)^{-1} = B^{-1}A^{-1}.$$

(c) $(aA)^{-1}(aA) = I \quad \Rightarrow$

$$\Rightarrow \quad (aA)^{-1}a = A^{-1} \quad \Rightarrow \quad (aA)^{-1} = a^{-1}A^{-1}.$$

d. Here we firstly, write down the above-mentioned equality and then consider the same equality, written in terms of the tensor components.

$$I = (A^T)^{-1}A^T \quad \Rightarrow$$

$$I_{ik} = (A^T)_{ij}^{-1}A_{jk}^T \quad \text{next, we transpose the cofactors}$$

$$= ((A^T)^{-1})_{ji}^T A_{kj} \quad \text{and change their order}$$

$$= A_{kj}((A^T)^{-1})_{ji}^T.$$

Thus, $I = A((A^T)^{-1})^T$

Then multiplying both sides by A^{-1}, we obtain

$$A^{-1} = ((A^T)^{-1})^T.$$

Finally, another transposition of both sides gives

$$(A^{-1})^T = (A^T)^{-1}.$$

3. Calculate theinner product of tensors

$$A{:}D, \quad A{:}W, \quad W{:}D,$$

where A is a tensor of the general form, while $D = D^T$ and $W = -W^T$ are symmetric and antisymmetric tensors, respectively. Show that the following equalities are valid.

a. $A{:}D = \frac{1}{2}(A + A^T){:}D$, *i.e.*, in the inner product of a general tensor A with a symmetric tensor D only the symmetric part of the tensor A is involved.

b. $A{:}W = \frac{1}{2})A - A^T({:}W$, similarly, in the inner product of a general tensor A with an antisymmetric tensor W only the antisymmetric part of the tensor A is involved.

c. $W : D = 0$ the inner product of an arbitrary symmetric tensor with an arbitrary antisymmetric tensor is equal to zero.

Solutions:

a. For any two tensors A and B the equality $A : B = \mathrm{tr}(AB)$ holds. Let a_{ij}, $d_{ij} = d_{ji}$ and $w_{ij} = -w_{ji}$ are components of the tensors A , D and W . Since, D is symmetric we may write

$$D = \frac{1}{2}(D + D^{T}).$$

Then the contraction of tensors A and D and its trace are as follows (note the order of indices)

$$(AD)_{ij} = \frac{1}{2}(A(D + D^{T}))_{ij} = \frac{1}{2}(a_{ik}d_{kj} + a_{ik}d_{jk}),$$

$$tr(AD) = \frac{1}{2}tr(AD) + \frac{1}{2}tr(AD^{T}) = \frac{1}{2}(a_{ik}d_{ki} + a_{ik}d_{ik}).$$

In the last expression, all indices are dummy. To find out, which tensor the tensor D is contracted with, it is necessary to factor it out. For this purpose we rename indices in the second term and transpose the tensor A (*i.e.*, interchange indices of its components) so as to be able to write the second term as a contraction.

$$tr(AD) = \frac{1}{2}(a_{ik}d_{ki} + a_{ki}d_{ki}) = \frac{1}{2}(a_{ik}d_{ki} + a_{ik}^{T}d_{ki}) = \frac{1}{2}(A:D + A^{T}:D).$$

Thus,

$$A:D = tr(AD) = \frac{1}{2}(A + A^{T}):D.$$

The validity of the remaining equalities is proved similarly.

4. Orthogonal Tensors

a. Show that tensors R_1, R_2 and R_3, with matrices

$$\begin{pmatrix} \cos\alpha & -\sin\alpha & 0 \\ \sin\alpha & \cos\alpha & 0 \\ 0 & 0 & 1 \end{pmatrix}, \begin{pmatrix} \cos\beta & 0 & -\sin\beta \\ 0 & 1 & 0 \\ \sin\beta & 0 & \cos\beta \end{pmatrix}, \begin{pmatrix} 1 & 0 & 0 \\ 0 & \cos\gamma & -\sin\gamma \\ 0 & \sin\gamma & \cos\gamma \end{pmatrix}$$

(with respect to the Cartesian basis), are orthogonal, *i.e.*, they do not change the length of a vector and each inverse matrix is equal to transposed. Find a geometric interpretation for the action of these tensors.

b. Show that arbitrary orthogonal (2 × 2) matrix may be represented as

$$\begin{pmatrix} \cos\phi & \pm\sin\phi \\ \sin\phi & \mp\cos\phi \end{pmatrix}$$

for a certain value of ϕ.

Solutions:

a. Let (a_1, a_2, a_3) are components of a vector $\vec{a}$ with respect to the same Cartesian basis. Firstly, compute $|\vec{a}|^2 = (\vec{a}, \vec{a})$ and $|R_i\vec{a}|^2 = (R_i\vec{a}, R_i\vec{a})$ for $i = 1, 2, 3$ and show that $|\vec{a}|^2 = |R_i\vec{a}|^2$. Next, show that the contraction $R_i R_i{}^T$ is equal to I for $i = 1, 2, 3$.

b. Let $\vec{a} = (a_1, a_2)$ and $R = \begin{pmatrix} r_{11} & r_{12} \\ r_{21} & r_{22} \end{pmatrix}$ with respect to the Cartesian basis. Firstly, we find r_{ij} such that $|\vec{a}| = |R\vec{a}|$ or $|\vec{a}|^2 = |R\vec{a}|^2$. In component form the latter equality is as follows

$$a_1^2 + a_2^2 = (r_{11}^2 + r_{21}^2)a_1^2 + (r_{12}^2 + r_{22}^2)a_2^2 + \\ +2(r_{11}r_{12} + r_{21}r_{22})a_1a_2,$$

which means that components r_{ij} satisfy the system of equations

$$r_{11}^2 + r_{21}^2 = 1, \tag{*1}$$

$$r_{12}^2 + r_{22}^2 = 1, \tag{*2}$$

$$r_{11}r_{12} + r_{21}r_{22} = 0. \tag{*3}$$

The equations (*3) and (*1) give

$$r_{11}^2 \overset{(*3)}{=} \left(\frac{r_{21}r_{22}}{r_{12}}\right)^2 r_{11}^2 \overset{(*1)}{=} 1 - r_{21}^2 \Rightarrow$$

$$\Rightarrow r_{12}^2 = r_{21}^2(r_{22}^2 + r_{12}^2) \overset{(*2)}{\Rightarrow} \quad r_{12}^2 = r_{21}^2.$$

At the same time the equations $((*3))$ and $((*2))$ imply

$$r_{12}^2 \overset{(*3)}{=} \left(\frac{r_{21}r_{22}}{r_{11}}\right)^2 r_{12}^2 \overset{(*2)}{=} 1 - r_{22}^2 \Rightarrow$$

$$\Rightarrow r_{11}^2 = r_{22}^2(r_{21}^2 + r_{11}^2) \overset{(*1)}{\Rightarrow} r_{11}^2 = r_{22}^2.$$

Besides, the equation $(*3)$ asserts that ratios $\frac{r_{12}}{r_{21}}$ and $\frac{r_{22}}{r_{11}}$ are of opposite signs and therefore $r_{12} = \pm r_{21}$ and $r_{11} = \mp r_{22}$. Thus, the diagonal elements in R are equal up to a sign. Now, since R is assumed to be orthogonal, it should possess the known property of orthogonal tensors $R^{-1} = R^T$, and its orthogonality may easily be checked *via* calculating the contraction RR^T. The result should be the unit tensor. Finally, the equations $(*1)$ and $(*2)$ shows that $|r_{ij}| \leq 1$. This allows interpretation of components of the tensor R as trigonometric functions $\sin\phi$ and $\cos\phi$. In this case the argument φ is understood as the angle between vectors $\vec{a}$ and $R\vec{a}$. As a result, we have

$$R = \begin{pmatrix} \cos\phi & \sin\phi \\ \pm\sin\phi & \mp\cos\phi \end{pmatrix} \quad or \quad R = \begin{pmatrix} \cos\phi & \pm\sin\phi \\ \sin\phi & \mp\cos\phi \end{pmatrix}.$$

5. Determine the vorticity of the flow. In the following descriptions of problems it is necessary to 1) formalize each problem and 2) choose an appropriate coordinate system (the location of its origin and its orientation) and vectors of the coordinate basis such that the problem gains the most simple form (directions and symmetries inherent to the problem should be used).

a. A stationary plane-parallel shear flow along an infinite flat solid surface. Consider linear, quadratic and logarithmic shear, as a function of the distance from the plate.

b. An elementary vortex: fluid rotates around an axis with time-independent rate. Consider rotation with an angular velocity 1) independent of the distance from the axis of rotation; 2) increases linearly with distance from the axis, and 3) inversely proportional to the distance.

Solutions:

a. Let's choose the Cartesian coordinate system as the most simple. We place the origin on the surface because this makes it motionless with respect to the coordinate system. Then, in order to null the two of the three components of the velocity vector with respect to the coordinate basis, we direct one of the coordinate axes (*e.g., z*) normal to the surface, and the other (say, *x*) along the velocity of the flow. The direction of the third axis, *y*, is inessential. In this case the velocity vector has only one non-zero component $\vec{v} = (0, v, 0)$, which is a function of only one variable $v = v(z)$, and the vorticity vector is as follows

$$\vec{\omega} = \nabla \times \vec{v} = \begin{vmatrix} \vec{e}_x & \vec{e}_y & \vec{e}_z \\ \partial_x & \partial_y & \partial_z \\ 0 & v & 0 \end{vmatrix} = -(\partial_z v)\vec{e}_x.$$

Now we may consider linear, quadratic and logarithmic shear, *i.e.*,

$$v = \begin{cases} \alpha z, \\ \alpha z^2, \\ \alpha \ln(z+1), \end{cases} \quad \alpha = const \in \mathbb{R}^1;$$

b. Let's choose the cylindrical coordinates system and write down the velocity of the flow in the form $\vec{v} = (v_r, rv_\varphi, v_z) = (0, rv_\varphi(r), 0)$. Next, we compute the $\vec{\omega}$ and consider following dependencies

$$v_\phi = \begin{cases} \alpha, \\ \alpha r, \\ \dfrac{\alpha}{r}, \end{cases} \quad \alpha = const \in \mathbb{R}^1;$$

16.3. WORKING WITH COMPONENTS AND THE SUMMATION RULE

1. **Auxiliary expressions**. Let f is an arbitrary scalar and let $\vec{v} = (v_1, v_2, v_3)$ and $\nabla = (\partial_{x1}, \partial_{x2}, \partial_{x3})$ with respect to the Cartesian coordinate basis.

a. Calculate:
   - div rot $\vec{v} = (\nabla, \nabla \times \vec{v}) = ?$
   - div grad $f = (\nabla, \nabla f) = ?$
   - rot grad $f = \nabla \times \nabla f = ?$

b. Show that the equality $(\vec{v}, \nabla)\vec{v} = \frac{1}{2}\nabla|\vec{v}|^2 - \vec{v} \times (\nabla \times \vec{v})$ holds. Write down the equality using corresponding components and simplify both sides of the expression.

2. Component form of equations.

a. Show that both forms of the continuity equation

$$d_t\rho + \rho(\nabla, \vec{v}) = 0 \quad \Leftrightarrow \quad \partial_t\rho + (\nabla, \rho\vec{v}) = 0.$$

may be derived from one another.

b. Show that the Euler equation

$$\rho d_t\vec{v} = -\nabla p + \rho\vec{b},$$

being written down in the coordinate form, is equivalent to the system of differential equations, describing the evolution of components of the velocity vector.

c. Using the continuity equation write down the left side of the Euler equations in the so-called divergent form:

$$\rho d_t v_k = \partial_t(\rho v_k) + (\nabla, \rho v_k \vec{v}).$$

Solutions:

a. Using the Euler's formula (3.10) we rewrite the total derivative and unite the last two terms

$$\begin{aligned}
0 = d_t\rho + \rho(\nabla, \vec{v}) &= d_t\rho + \rho\,\partial_{x_i}v_i \\
&= \partial_t\rho + v_i\,\partial_{x_i}\rho + \rho\,\partial_{x_i}v_i \\
&= \partial_t\rho + \partial_{x_i}(\rho v_i) = \partial_t\rho + (\nabla, \rho\vec{v}).
\end{aligned}$$

b. Assuming the body force is equal to the Earth's gravity, we choose the Cartesian coordinate system and corresponding Cartesian basis such that the vector $\vec{b}$ have only one non-zero component (*e.g.*, $b_3 = \pm g$). Using the Kronecker delta we may write down the specific body force density in the form $\vec{b} = \pm g\delta_{k3}\vec{e}_k$. Next, we consider the differential operator $\nabla \equiv \vec{e}_i\partial_{x_i}$ and compute ∇p. Then, applying the Euler's formula to the left-hand side, we write down the equation of motion in the form:

$$(\rho\,\partial_t v_k)\vec{e}_k + (\rho v_i\,\partial_{x_i}v_k)\vec{e}_k = (-\partial_{x_k}p)\vec{e}_k + (\pm\rho g\delta_{k3})\vec{e}_k.$$

Factoring out the basis vector we obtain the required system of equations (here k is a free index, which numbers the equations in the system)

$$\rho \, \partial_t v_k + \rho v_i \, \partial_{x_i} v_k = - \partial_{x_k} p \pm \rho g \delta_{k3}.$$

c. Summing the left-hand side of the Euler equations, written down for the k-th component of velocity vector, with the continuity equation, multiplied by the k-th component of velocity, we shall obtain the desired result

$$\rho d_t v_k = \rho \, \partial_t v_k + \rho v_i \, \partial_{x_i} v_k + v_k \left(\partial_t \rho + \partial_{x_i} (\rho v_i) \right)$$
$$= \partial_t \rho v_k + \partial_{x_i} (\rho v_k v_i) = \partial_t \rho v_k + (\nabla, \rho v_k \vec{v}).$$

16.4. HYDROSTATICS AND THE PERFECT FLUID

Making the most of the conditions of the problem, choose an appropriate coordinate system (including the location of its origin, its orientation), basis vectors, model of the fluid. Next, formalize a verbal description of the problem and solve it.

1. (Mase [15]) A wide vessel filled with a fluid of constant density, moves with constant acceleration $\vec{a}$ in a gravity field. Determine the shape of the free surface if the surface pressure is known, constant and equal to p_0.
2. (Landau & Lifshits [16]) A wide cylindrical vessel filled with a fluid of constant density, rotates about its (vertical) axis with a constant angular velocity in a gravity field. Determine the shape of the free surface if the surface pressure is known, constant and equal to p_0.
3. (Mase [15]) A fluid with the equation of state $p = \lambda \rho^k$, where $\lambda, k = \text{const}$ flows from a large closed tank through a smooth thin pipe. The pressure in the tank is N times the atmospheric pressure. Determine the velocity of the emerging fluid.
4. (Mase [15]) Given a velocity potential $\varphi = Ax + B \frac{x}{r^2}$, where $r^2 = x^2 + y^2$. Determine the stream function ψ.

Solutions:

1. Since a vessel is wide, the shape of the surface of the fluid is determined by the balance of the contact and body forces, and distortion due to wetting the walls may be neglected. Fluid is moving together with the vessel and there is no motion relative to the vessel. Therefore, in the coordinate system moving with the vessel, the hydrostatic system of equations may be used. We choose the Cartesian coordinate system ($x_1 = x, x_2 = y, x_3 = z$), place the origin (the point o) on the fluid surface, direct the z-axis upwards, and orient the y-axis such that the vector $\vec{a}$ lies in the plane of yoz. At each point, we select a coordinate basis $\{\vec{e}_i\} = \{\vec{e}_x, \vec{e}_y, \vec{e}_z\}$. With respect to this basis the gravity acceleration $\vec{b}$ and

the constant acceleration $\vec{a}$ have the following components $\vec{b} = (0, 0, -g)$ and $\vec{a}$ $= (0, a_y, a_z)$. Hydrostatic system of equations in this case is as follows

$$\partial_x p = 0,$$

$$\partial_y p = \rho a_y,$$

$$\partial_z p = \rho(a_z - g).$$

Integrating the first equation, we get $p = p(y, z)$. The solution of the second equation is the function $p = \rho a_y y + C_1 (z)$. Substituting this result into the third equation and integrating we obtain: $C_1 = \rho(a_z - g)z + C_2$. So, the general solution for the pressure is as follows:

$$p = \rho a_y y + \rho(a_z - g)z + C_2$$

The constant C_2 is determined using the boundary condition $p|_{z=0} = p_0$ in the origin ($x = z = 0$). This gives $C_2 = p_0$ and

$$p = \rho a_y y + \rho(a_z - g)z + p_0.$$

Since the surface of the liquid is determined by the condition $p = p_0$, we get the equation, which describe this surface: $z = \dfrac{a_y}{g - a_z} y$, i.e. the surface consists of points with coordinates $\left(x, y, \dfrac{a_y}{g - a_z} y\right)$.

2. Here we also choose the Cartesian coordinates (x, y, z), place the origin at the surface of the fluid, and arrange axes such that the angular velocity vector has a single non-zero component $\Omega = d_t \phi$. where phi is time dependent angle between the x-axis and constantly changing direction to the moving point. Let r is a distance from the point with coordinates (x, y) to the axis of rotation. Then the components of the velocity vector of this point are as follows

$$u = d_t x = d_t(r\cos\phi) = -y\Omega,$$
$$v = d_t y = d_t(r\sin\phi) = x\Omega,$$
$$w = 0.$$

Substituting these velocity components in the Euler equations of motion, we obtain three scalar equations for the pressure:

$$\begin{cases} x\Omega^2 = \dfrac{1}{\rho}\partial_x p & \Rightarrow & p = \dfrac{1}{2}\rho\Omega^2 x^2 + C_1(y, z), \\[2mm] y\Omega^2 = \dfrac{1}{\rho}\partial_y p & \Rightarrow & y\Omega^2 = \dfrac{1}{\rho}\partial_y C_1 & \Rightarrow & C_1 = \dfrac{1}{2}\rho\Omega^2 y^2 + C_2(z), \\[2mm] 0 = \dfrac{1}{\rho}\partial_z p + g & \Rightarrow & \dfrac{1}{\rho}\partial_z C_2 + g = 0 & \Rightarrow & C_2 = -\rho g z + C_3. \end{cases}$$

Since everywhere on the surface $p = p_0$, the condition $p|_{x = y = z = 0} = p_0$ allows determining of the constant $C_3 = p_0$. The surface itself is determined by the condition $p = p_0$, which give us the equation, describing this surface $z = \dfrac{\Omega^2}{2g}(x^2 + y^2)$. Hence, the surface is a paraboloid of revolution.

3. Since, the flow is stationary, we solve the problem using the Bernoulli integral H. Inside the tank ($\vec{v}_i = 0$, $p_i = N$, $\Phi_i = \Phi$) we have:

$$H_i = \frac{1}{2}|\vec{v}_i|^2 + P_i + \Phi_i = P_i + \Phi.$$

Outside the tank ($\vec{v}_o$, $p_o = 1$, $\Phi_o = \Phi$) the Bernoulli integral is equal to:
$$H_o = \frac{1}{2}|\vec{v}_o|^2 + P_o + \Phi_o = \frac{1}{2}|v_o|^2 + P_o + \Phi.$$

The function $P = \int_{p_i}^{p} \frac{1}{\rho} dp$ may be calculated using the equation of state. Firstly, we find $dp = \lambda k \rho^{k-1} d\rho$ and further:

$$P = k\lambda \int_{\rho(p_i)}^{\rho(p)} \rho^{k-2} d\rho = \frac{k\lambda}{k-1} \rho^{k-1}\Big|_{\rho(p_i)}^{\rho(p)} =$$
$$= \left(\frac{1}{\lambda}\right)\frac{k}{k-1}\left(\frac{p}{\rho} - \frac{p_i}{\rho_i}\right).$$

As a result we obtain two values of P inside and outside of the tank $P_i = 0$, $P_o = \frac{k}{k-1}\left(\frac{1}{\rho_o} - \frac{N}{\rho_i}\right)$. Since the values of the Bernoulli integral inside and outside the tank should be equal one another $H_i = H_o$, we have

$$|\vec{v}_o|^2 = -2P_o = \frac{2k}{k-1}\frac{1}{\rho_o}\left(N\frac{\rho_o}{\rho_i} - 1\right),$$

or using the equation of state, we get $\frac{\rho_o}{\rho_i} = N^{-\frac{1}{k}}$, and finally obtain

$$|\vec{v}_o|^2 = \frac{2k}{k-1}\frac{1}{\rho_o}\left(N^{1-\frac{1}{k}} - 1\right).$$

4. Using definitions of the stream function and the velocity potential we may construct the following system of equations

$$\begin{aligned}\partial_x\varphi &= \partial_y\psi, \\ \partial_y\varphi &= -\partial_x\psi.\end{aligned} \tag{*4}$$

Differentiating the given expression for potential φ with respect to x, and substituting the result to the first equation of the system, we find

$$\partial_y \psi = A - B \frac{x^2 - y^2}{(x^2 + y^2)^2}.$$

The solution of the latter equation contains the unknown function $C_1(x)$:

$$\psi = Ay - B \frac{y}{x^2 + y^2} + C_1(x).$$

Differentiating this expression with respect to x and substituting the result into the second equation of (*4), we get

$$\partial_y \varphi = -2B \frac{xy}{(x^2 + y^2)^2} - d_x C_1(x).$$

The solution of this equation gives us the expression for potential

$$\varphi = B \frac{x}{x^2 + y^2} - y d_x C_1 + C_2(x),$$

where $C_2(x)$ is a one more "constant" of integration. Comparing the obtained expression with conditions of the problem, we derive the equation which allows finding of the function $C_1(x)$

$$d_x C_1(x) = \frac{1}{y}(-Ax + C_2(x)). \qquad \text{(*5)}$$

Since C_1 is a function of one argument x, the equation (*5) by necessity implies $C_2(x)$ = Ax and $C_1 = C =$ const. Thus, using the obtained expression for C_1, we finally have

$$\psi = -B \frac{y}{x^2 + y^2} + Ay + C.$$

16.5. VISCOUS FLUID

1. **The equation of motion of a viscous fluid.**
   Write down in the component form

a. the Navier-Stokes equation

$$\rho d_t \vec{v} = -\nabla p + \mu \Delta \vec{v} + \rho \vec{b},$$

i.e., derive a system of differential equations describing the evolution of components of the velocity vector;

b. the internal energy density balance equation of a viscous fluid:

$$d_t \varepsilon = -p d_t \frac{1}{\rho} + 2\nu D : D + \frac{1}{\rho}(\nabla, \vec{h}) + s.$$

2. **Parallel flows: the exact solutions of the Navier-Stokes equations.**
   A liquid layer of thickness h is located between two infinite parallel planes moving with a constant velocity U relative to each other and parallel to themselves. Describe the steady flow of fluid in the absence of gravity and in the presence of a constant pressure gradient co-directed with the velocity U. Formulate the boundary value problem, solve it and calculate the flow rate.

   a. The Couette[1] flow: pressure gradient is absent.
   b. The Poiseuille[2] flow: $U = 0$.
   c. The generalized Couette flow.

3. (Landau & Lifshits [16]) A liquid layer of thickness h is bounded from below by a fixed plane inclined at an angle α to the horizon. Determine the steady flow of a viscous fluid due to gravity, if the surface pressure is $p_0 = $ const.
4. Write the equation of the heat transfer in the non-dimensional form.

Solutions:

2. Selecting suitable system of coordinates (x, y, z) we reduce the system of the Navier-Stokes equations to a single equation

$$\rho \, \partial_t u = -\partial_x p + \mu \partial_{zz} u$$

   for the only non-zero component of the velocity $\vec{v}(z) = (u(z), 0, 0)$. Further we formulate the boundary value problem, solve it and calculate the flow rate according to the formula

$$D \equiv \int_0^H u(z)dz.$$

   a. Boundary conditions for the Couette flow $(\partial_x p = 0)$:

$$\partial_{zz} u = 0, \qquad u|_0 = 0, \quad u|_h = U.$$

   Answer:

$$u = U\frac{z}{h}, \qquad D = \frac{1}{2}Uh.$$

   b. Boundary conditions for the Poiseuille flow $(U = 0, \quad \partial_x p = $ const $\neq 0)$:

$$\mu \partial_{zz} u = \partial_x p, \qquad u|_0 = 0, \quad u|_h = 0.$$

   Answer:

$$u = -\frac{(h-z)z}{2\mu}\partial_x p, \qquad D = -\frac{h^3}{12\mu}\partial_x p.$$

c. Boundary conditions for the generalized Couette flow ($U = $ const $\neq 0$, $\partial_x p = $ const $\neq 0$):

$$\mu\, \partial_{zz} u = \partial_x p, \qquad u|_0 = 0, \quad u|_h = U.$$

Answer:

$$u = U\frac{z}{h} - \frac{(h-z)z}{2\mu}\partial_x p, \qquad D = \frac{1}{2}Uh - \frac{h^3}{12\mu}\partial_x p.$$

3. We place the origin of the Cartesian coordinate system on a fixed plane (x-axis is directed perpendicular and y-axis is parallel to the horizon, z-axis is directed upward). The task is reduced to finding the flow along a horizontal border under the action of a fixed body force with specific density ($g \cos \alpha$, 0, $g \sin \alpha$). The equations, describing the velocity and pressure distributions are as follows:

$$\partial_{zz} u = -\frac{g}{\nu}\sin\alpha,$$
$$\partial_z p = -\rho g \cos\alpha.$$

Further, we consider the boundary conditions (no-slip condition at the lower boundary, zero gradient of the u-component and constant pressure p_0 at the upper boundary):

$$u|_{z=0} = 0,$$
$$\partial_z u|_{z=h} = 0,$$
$$p|_{z=h} = p_0.$$

Integrating the system, we find

$$p = p_0 + (h - z)\rho g \cos\alpha,$$
$$u = \frac{g}{2\nu}z(2h - z)\sin\alpha.$$

Note, that in case of vanishing viscosity $\nu \to 0$ (the fluid is perfect) the velocity tends to infinity $\lim_{\nu \to 0} u = \infty$.

16.6. THE HEAT EQUATION

1. Using the general equation of the heat conduction of a moving fluid

$$d_t T = \partial_t T + (\vec{v}, \nabla T) = \kappa \Delta T + s_T$$

derive an equation describing a particular case: the medium is one-dimensional, the temperature distribution is stationary, the velocity of the medium is constant, there are no external sources of heat.

2. Consider an interval $[a, b]$ of the one-dimensional medium (see the above-mentioned particular case), moving with the flow. Find the temperature

distribution on this interval for the following boundary conditions:

a. temperature values T_a and T_b at endpoints of the interval are specified,
b. the following values are specified: the temperature $T|_{x=a} = T_a$ and the flux $q_b = (\lambda \partial_x T)_{x=b}$,
c. the temperature T_a at one end and the average temperature on the whole interval

$$\bar{T} = \frac{1}{b-a} \int_a^b T(x) dx$$

   are specified,

d. The value of the flux q_a at one end of the interval and the average temperature $\bar{T}$ on the whole interval are specified.

3. Let at both ends of the interval the values of the flux q_a and q_b are specified. Find additional condition which is necessary for the existence of the solution of the problem, and show that this solution is not unique.

Solutions:

1. The mentioned limitations lead to following simplifications of the equation:

a. stationary temperature distribution $\Rightarrow$ $\partial_t T \equiv 0$,
b. no external sources $\Rightarrow$ $s_T \equiv 0$,
c. one-dimensional media $\Rightarrow$ a point is described using only one coordinate, and a velocity vector has a single component,
d. velocity of the medium is constant $\Rightarrow$ a moving coordinate system, with respect to which the medium will be at rest, may be selected.

   If we denote the coordinate by x, the equation will be written in the form

$$\partial_{xx} T = 0.$$

2. To solve the equation, we introduce an auxiliary function f such that

$$f(x) \equiv \partial_x T.$$

Now the original equation may be rewritten as $\partial_x f = 0$ and may easily be integrated. The solution is $f(x) = C_1 = \text{const}$. Next we use the definition of the function f as the equation to find the unknown T

$$\partial_x T = C_1.$$

This equation is also easy to integrate. The result is $T = C_1 x + C_2$. It remains to determine both constants using the boundary conditions. For example, in the latter case we have:

- $$q_a = (\lambda \partial_x T)_{x=a} = \lambda C_1 \quad \Rightarrow \quad C_1 = \frac{q_a}{\lambda} \quad \Rightarrow \quad T = \frac{q_a}{\lambda} x + C_2,$$

- $$\overline{T} = \frac{1}{b-a} \int_a^b T(x) dx = \frac{1}{b-a} \int_a^b \left(\frac{q_a}{\lambda} x + C_2 \right) dx = \frac{q_a}{\lambda} \cdot \frac{b+a}{2} + C_2 \quad \Rightarrow \quad C_2 = \overline{T} - \frac{q_a}{\lambda} \cdot \frac{b+a}{2}.$$

Thus,

$$T = \overline{T} + \frac{q_a}{\lambda} \left(x - \frac{b+a}{2} \right).$$

3. From the definition of the heat flux $q = (\lambda \partial_x T)$ and the solution of the equation, it follows that $q = C_1$ everywhere on the interval $[a, b]$. Thus, specification of different values of the flux at both ends of the interval is contradictory. In other words, there is no solution with such values of the flux at the boundaries. Solution exists only when the values of flux at the ends of the interval are the same. However, these boundary conditions are equivalent to a single condition. The absence of the second condition leads to the fact that an infinite number of solutions exists.

16.7. TURBULENT FLUID

1. Show that the following properties of the averaging operator
   - linearity: $\overline{af + bh} = a\overline{f} + b\overline{h}$, a, b = const,
   - continuity: $\overline{\lim_{n \to \infty} f_n} = \lim_{n \to \infty} \overline{f}_n$,
   - projector: $\overline{\overline{f}} = \overline{f}$.

imply:

a. $\overline{a} = a$, a = const,

b. $\overline{af} = a\overline{f}$, a = const,

c. $\overline{\partial_{x_i} f} = \partial_{x_i} \overline{f}$,

d. $\overline{f + g} = \overline{f} + \overline{g}$,

e. $\overline{\overline{f} \, g} = \overline{f} \, \overline{g}$.

Solutions:

a. since a = const, the fluctuation $a' = 0$ and $a = \overline{a} + a' = \overline{a}$; or since, the averaging operator is a projector, we have $\overline{\overline{a} + a'} = \overline{\overline{a}} = \overline{a} = a$;

b. the equality holds due to linearity of the averaging operator;

c. the following sequence of equalities

$$\overline{\partial_{x_i} f} \overset{(a)}{=} \overline{\lim_{\Delta x_i \to 0} \frac{\Delta f}{\Delta x_i}} \overset{(b)}{=} \lim_{\Delta x_i \to 0} \frac{\overline{\Delta f}}{\Delta x_i} \overset{(c)}{=} \lim_{\Delta x_i \to 0} \frac{\Delta \bar{f}}{\Delta x_i} \overset{(a)}{=} \partial_{x_i} \bar{f}$$

hold due to definition of the derivative (a), continuity of the averaging operator (b) and its linearity (c).

d. equality holds due to linearity of the averaging operator;

e. equality is also due to linearity of the averaging operator.

2. Show that the equality $\overline{\bar{f} g} = \bar{f} \, \bar{g}$, where g takes values $g = 1$, $\bar{h}$ and h' is equivalent to the following three equalities:

a. $\bar{\bar{f}} = \bar{f}$,

b. $\overline{\bar{f} \bar{h}} = \bar{f} \, \bar{h}$,

c. $\overline{h'} = 0$.

3. Let the turbulence is generated exclusively by velocity fluctuations. Prove that

a. the averaged continuity equation has the form

$$\partial_t \rho + (\nabla, \rho \bar{\vec{v}}) = 0.$$

b. velocity fluctuations satisfy the equation

$$(\nabla, \rho \vec{v}') = 0.$$

Solutions:

a. We write the velocity in the form $\vec{v} = \bar{\vec{v}} + \vec{v}'$, substitute in the continuity equation and average.

$$\overline{\partial_t \rho + (\nabla, \rho(\bar{\vec{v}} + \vec{v}'))} = \partial_t \rho + (\nabla, \rho \bar{\vec{v}}) + \underbrace{(\nabla, \rho \, \overline{\vec{v}'})}_{=0} =$$

$$= \partial_t \rho + (\nabla, \rho \bar{\vec{v}}) = 0.$$

The problem 2c implies $\overline{\vec{v}'} = 0$, and this gives $\partial_t \rho + (\nabla, \rho \bar{\vec{v}}) = 0$.

b. We write the velocity as in the previous item and substitute in the continuity equation

$$\partial_t \rho + (\nabla, \rho(\bar{\vec{v}} + \vec{v}')) = \underbrace{\partial_t \rho + (\nabla, \rho\bar{\vec{v}})}_{=0} + (\nabla, \rho\vec{v}') = 0.$$

In the task (a) it has been shown that the first two terms are equal to zero (the averaged continuity equation). Hence, $(\nabla, \rho\vec{v}') = 0$.

NOTES

[1]Couette Maurice Marie Alfred (1858–1943), a French hydromechanics.

[2]Poiseuille Jean Louis Marie (1799-1869), a French physician and physicist.

List of Notations

NB! numbers indicate pages of the first appearance of corresponding notations.

LATIN LETTERS

A, B, ...   — tensors of the second rank, 54

$\mathcal{B}, \mathcal{C}, \ldots$ — bodies, 3

$\mathcal{B}^e$ — exterior of the body $\mathcal{B}$, 14

$\vec{b}$ — specific body force density, 85

c_p — heat capacity at constant pressure, 145

D   — deformation rate tensor, 70

d_t — total time derivative, 48

∂_t — partial time derivative, 48

E — internal energy, 101

$\vec{e}_i$ — basis vector, 8

F   — deformation gradient, 65

$\mathbf{f}_{\mathcal{B}}(\mathcal{C}, t)$ — force applied by $\mathcal{C}$ to $\mathcal{B}$ at time t, 14

$\mathbf{f}_C$ — surface force, 85

$\mathbf{f}_B$ — body force, 85

G   — velocity gradient, 69

H — Bernoulli integral, 115

$\vec{h}$ — heat flux density, 104

I   — unit tensor, 59

J — Jacobian of coordinate transformation, 29

K — kinetic energy, 96

k — kinetic energy density, 96

$M(\mathcal{B})$ — mass of the body $\mathcal{B}$, 12

$\mathbf{m}(t, \mathcal{B})$ — momentum of the body $\mathcal{B}$, 82

$\vec{n}$ — outward normal vector to a point of a surface, 80

O — zero tensor, 59

P_t — space of places, 3

$P(p) = \int \frac{1}{\rho} dp$ — 114

$\mathcal{P}, \mathcal{P}_k$ — parts of a body, 22

p — pressure, 91

Q — heating rate, 103

Q_B — radiation heating rate, 103

Q_C — surface heating rate, 103

q — heat flux, 103

R — orthogonal tensor, 62; Reynolds stress tensor, 165

$\mathbb{R}^{0+}$ — set of non-negative real numbers, 12

$\mathbb{R}^1, \mathbb{R}^3, \mathbb{R}^4$ — real coordinate spaces of 1, 3 and 4 dimensions, 5

S — surface bounding a certain volume, 85

s — specific heat flux due to radiation, 94

T — temperature, 145

T — stress tensor, 88

T' — viscous stress tensor, 122

t — time, parameter of a world-line, 3

$\vec{u} = d_t \lambda$ — vector tangent to a world line, 7

U — potential energy, 101

V — volume of a body configuration, 22

$\vec{v}$ — velocity vector, 10

v_* — friction velocity, 181

W — power, 98

$\mathcal{W}$ — space of events, 3

W — spin, 70

$X, Y, \ldots$ — points of a body, 3

$\mathbf{X} = (X_1, X_2, X_3)$ — Lagrangian coordinates of a point of a body, 26

$\mathbf{x} = (x_1, x_2, x_3)$ — coordinates of a place, 5; Euler coordinates, 27

$\mathbf{x}(t)$ — parameterized curve, 9

z_0 — roughness parameter, 185

GREEK LETTERS

Δ — increment, 19; Laplacian, 116

δ_{ij} — Kronecker delta, 59

ε — internal energy density, 101

ε^{ijk} — Levi-Civita symbol, 74

κ — thermal diffusivity, 146

$\kappa(\mathcal{B})$ — reference configuration of a body $\mathcal{B}$, 26

λ — thermal conductivity, 146

$\lambda(t, X)$ — world-line of a point X, 3

$\lambda(t, \mathcal{B})$ — world-tube of a body $\mathcal{B}$, 3

μ — dynamic viscosity, 136

ν — kinematic viscosity, 138

Π — integral parameter of a body, 22

π — density of an integral parameter, 23

ρ — mass density, 23

Σ — power of a source of an integral parameter, 30

σ — flux of an integral parameter, 30

$\vec{\tau}$ — surface force density, stress, 85

Φ — body force potential, 86

φ — velocity potential, 132

ϕ — frame of reference, 5

χ_κ — deformation, 27

$\chi(t, X)$ — trajectory of the point X, 9

$\chi(t, \mathcal{B})$ — configuration of the body $\mathcal{B}$ at time t, 9

ψ — stream function, 128

Ω — set of bodies with mass, 12

$\vec{\omega}$ — vorticity, 71

MATHEMATICAL SYMBOLS

$(\cdot)$ — place holder, 3

$\emptyset$ — empty set, 12

$\cup$ — union of sets, 12

$\cap$ — intersection of sets, 12

∇_k — covariant derivative in the direction of a vector $\vec{e}_k$, 48

∇ — nabla (Hamiltonian), 43

$(\cdot)^\mathsf{T}$ — transposition, 60

$(\vec{a}, \vec{b})$ — inner (scalar) product of vectors, 43, 57

$\mathsf{A} : \mathsf{B}$ — inner product of tensors, 57

$\det(\mathsf{A})$ — determinant of a square matrix, 74

$[a]$ — physical dimension of a quantity a, 151

Text Books for Further Reading

1. G. K. Batchelor, *An Introduction to Fluid Dynamics*. Cambridge University Press, Cambridge, UK, 2000.
2. S. Childress, *An Introduction to Theoretical Fluid Mechanics*, ser. Courant Lecture Notes in Mathematics, 19. A.M.S., Providence, 2009.
3. A. J. Chorin and J. E. Marsden, *A Mathematical Introduction to Fluid Mechanics*, 3rd ed. Springer-Verlag. New York, 1993.
4. N. E. Kochin, I. A. Kibel, and N. V. Rose, *Theoretical Hydromechanics*. Wiley-Interscience, 1964, vol. 1.
5. L. Landau and E. Lifshitz, *Fluid Mechanics*. Oxford: Pergamon, 1975.
6. R. E. Meyer, *Introduction to Mathematical Fluid Dynamics*. New York, Dover, 1982.
7. L. I. Sedov, *Mechanics of Continuous Media*. River Edge, NJ: World Scientific, 1997, vol. 1.
8. J. Serrin, "Mathematical principles of classical fluid mechanics," in *Handbuch der Physik, Bd VIII/1*, pp. 125–263, Springer Verlag, 1959.

References

[1]   C. Truesdell, *A First Course in Rational Continuum Mechanics*. Academic Press: New York, 1977.

[2]   W. Rudin, *Principles of Mathematical Analysis*. 3rd ed McGraw-Hill, 1976.

[3]   N.R. Lebovitz, Ed., *Fluid dynamics in astrophysics and geophysics*. 1983.

[4]   J. Serrin, *Mathematical principles of classical fluid mechanics*. 1959.
       [http://dx.doi.org/10.1007/978-3-642-45914-6_2]

[5]   B.F. Schutz, *Geometrical methods of mathematical physics*. Cambridge University Press: Cambridge, 1980.
       [http://dx.doi.org/10.1017/CBO9781139171540]

[6]   I. Gyarmati, *Non-equilibrium Thermodynamics. Field Theory and Variational Principles*. Springer: New York, 1970.
       [http://dx.doi.org/10.1007/978-3-642-51067-0]

[7]   G.I. Barenblatt, *Similarity, self-similarity, and intermediate asymptotics*. Consultants Bureau, Plenum Press, 1979.
       [http://dx.doi.org/10.1007/978-1-4615-8570-1]

[8]   A.S. Monin, and A.M. Yaglom, *Statistical Fluid Mechanics*, vol. 1. MIT Press: Cambridge, 1971.

[9]   J. Lumley, and H. Panofsky, *The structure of atmospheric turbulence*. Interscience Pulbishers, 1964.

[10]   H. Schlichting, *Boundary Layer Theory*. McGraw-Hill: New York, NY, 2004.

[11]   M. Belevich, "Causal description of non-relativistic dissipative fluid motion", *Acta Mech.,* vol. 161, pp. 65-80, 2003.
       [http://dx.doi.org/10.1007/s00707-002-0983-0]

[12]   M. Belevich, "Relationship between standard and causal fluid models", *Acta Mech.,* vol. 180, pp. 83-106, 2005.
       [http://dx.doi.org/10.1007/s00707-005-0262-y]

[13]   M. Belevich, "Thermodynamics from an observer's viewpoint (on the example of the viscous fluid)", *Contin. Mech. Thermodyn.,* vol. 26, no. 3, pp. 303-320, 2014.
       [http://dx.doi.org/10.1007/s00161-013-0303-z]

[14]   P. Glansdorff, and I. Prigogine, *Thermodynamic theory of structure, stability and fluctuations*. Wiley-Interscience: London, New York, 1971.

[15]   G.E. Mase, *Theory and Problems of Continuum Mechanics*. McGraw-Hill, 1970.

[16]   L. Landau, and E. Lifshitz, *Fluid Mechanics*. Pergamon: Oxford, 1975.

SUBJECT INDEX

A

additivity, 12
antisymmetric tensor, 60
averaging interval, 161
averaging, 159

B

balance equation of mechanical energy density, 100
balance equation of the total energy density, 104
balance relations, 30
baroclinic model, 108
barotropic model, 108
basic law of motion, 86
basic principle of dynamic, 83
basis vector, 8
Bernoulli equation, 115
Bernoulli integral, 115
Bernoulli's theorem, 116
body force density, 85
body force, 84
body heating, 103
body, 2
Boltzmann postulate, 90
boundary layer separation, 177
boundary surface, 22

C

Cauchy lemma, 88
Cauchy postulate, 88
Cauchy theorem, 88
causality principle, 30
center of mass, 82
Christoffel symbol, 50
coefficient of connexion, 50
coefficient of molecular diffusion, 147
coefficient of thermal expansion, 148
completely rough wall, 182
compressibility, 81
compressible fluid, 81

configuration, 9
conservation laws, 29
contact force density, 85
contact force, 84
contact heating, 103
continuity equation, 80
continuity hypothesis, 20
continuum, 20
contraction, 56
coordinate basis, 10
coordinates, 5
covariant derivative, 48
current configuration, 26

D

deformation gradient, 65
deformation rate tensor, 70
deformation, 27
density, 23
differential balance equation, 78
differential conservation law, 78
diffusion equation, 147
dimensional considerations, 181
displacement thickness, 173
dissipation of kinetic energy, 142
divergence, 58
dummy index, 42
dynamic similarity, 154
dynamic viscosity, 136

E

eddy viscosity, 165
eigenvalue problem, 62
eigenvalue, 62
eigenvector, 62
enthalpy, 145
entropy, 146
equation of motion in the Lamb form, 112
equation of motion, 90
equations of state, 106
Euler coordinates, 27
Euler equation, 92